OXFORD SERIES IN OPTICAL AND IMAGING SCIENCES

1. D. M. Lubman (ed.): *Lasers and mass spectrometry*
2. D. Sarid: *Scanning force microscopy with applications to electric, magnetic, and atomic forces*
3. A. B. Shvartsburg: *Non-linear pulses in integrated and waveguide optics*
4. C. J. Chen: *Introduction to scanning force microscopy*
5. D. Sarid: *Scanning force microscopy with applications to electric, magnetic, and atomic forces* Revised edition
6. S. Mukamell: *Principles of nonlinear optical spectroscopy*
7. Akira Hasegawa and Yuji Kodama: *Solitons in optical communications*
8. A. B. Shvartsburg: *Time-domain optics of ultrashort waveforms*

Time-domain optics of ultrashort waveforms

A. B. SHVARTSBURG
Central Design Bureau for Unique Instrumentation, Moscow

CLARENDON PRESS • OXFORD
1996

Oxford University Press, Walton Street, Oxford, OX2 6DP
Oxford New York
Athens Auckland Bangkok Bombay
Calcutta Cape Town Dar es Salaam Delhi
Florence Hong Kong Istanbul Karachi
Kuala Lumpur Madras Madrid Melbourne
Mexico City Nairobi Paris Singapore
Taipei Tokyo Toronto
and associated companies in
Berlin Ibadan

Oxford is a trade mark of Oxford University Press

Published in the United States
by Oxford University Press Inc., New York

A catalogue record for this book is available from the British Library

Library of Congress Cataloging in Publication Data

Shvartsburg, A. B.
Time-domain optics of ultrashort waveforms / A. B. Shvartsburg.
(Oxford series in optical and imaging sciences ; 8)
1. Electromagnetic waves–Transmission. 2. Quantum electrodynamics. 3. Time-domain analysis. I. Title. II. Series.
QC665.T7S5 1996 539.2–dc20 95–45402

ISBN 0 19 856509 7 (Hbk)

Typeset by Colset Pte Ltd, Singapore
Printed in Great Britain by
Bookcraft (Bath) Ltd
Midsomer Norton, Avon

PREFACE

Single pulses are coming

This book is devoted to the time-domain transient electrodynamics of dispersive and lossy media. This topic is associated with unusual interaction of ultrashort transient signals, containing only one or a few oscillations of an electromagnetic field with these media. Use of such single-cycle waveforms and nonsinusoidal EM fields excited by these waveforms in dielectric and conducting materials opens up new horizons in the nonstationary electrodynamics of continuous media. The current interest in these problems is threefold. First, progress in generation and detection of ultrashort transients in infrared optics and microwave physics has recently given rise to the creation of innovative methods for research on materials, such as transient spectroscopy, texture analysis or shock excitement of molecular vibrations. Second, the controlled shaping of solitary pulses broadens the possibilities of fractal communications, signal recording on a noise background and processing of ultrashort waveforms in real time. Third, the development of electrodynamics of single waveforms can provide optimized nonstationary transfer of energy and information through dispersive and absorbing media. The design of devices using these pulse phenomena has become the 'hot' topic of modern optoelectronics and radiophysics. This topic is in demand for science, communications and defence.

However, the physical fundamentals and the mathematical background of this new branch of electrodynamics are still insufficiently developed. The traditional analysis of pulse EM phenomena in continuous media is based on Fourier expansions of both fields and time-dependent electromagnetic parameters of the medium, e.g. current or polarization, with respect to the frequency or wavenumber. This approach is known to be a powerful tool for dealing with the long trains of slowly varying quasi-harmonic waves containing many field oscillations; such wave trains are radiated by continuous sources in conventional optics, radio technology and acoustics. In contrast, attempts to apply the travelling wave concept to the analysis of interactions of single ultrashort arbitrarily shaped waveforms with dispersive media are hampered for the following reasons:

1. A Fourier transform breaks a signal into continuously repeating components at various scales, so it is unsuitable for representing signals that change suddenly over time.
2. A short pulse, characterized by considerable spectral bandwidth, is known to spread rapidly in the course of propagation in a dispersive

medium. However, frequency-domain analysis of the controlled shaping of such rapidly distorting broadband waveforms at depth in a dispersive medium is difficult to achieve.

3. The signals incident on the dispersive interface may be structured on many different scales. Small-scale structure in one area of the signal is often unrelated to the structure in a different area. Attempts to use Fourier transforms for the determination of reflection and scattering coefficients of such signals result in complex integrals in the complex plane. Analytical calculation of these integrals may be considered sometimes as an art; meanwhile, numerical solutions still seem to be impractical. These solutions engender many problems, connected with numerical instabilities, interpolation error, accumulation of errors with frequency marching, etc. Considerable physical insight is lost through these solutions; moreover, one pays a penalty for this computational electromagnetics, owing to the increasing cost for computer time.

Thus the Fourier approach, which is a cornerstone for a multitude of topics in optics, radiophysics, and wave mechanics, proves to be inadequate for some problems in nonstationary electrodynamics. To develop an adequate approach to these problems, one should operate directly with Maxwell equations beyond the scope of Fourier representations. Such time-domain nonstationary electrodynamics is developed in this book.

Wide classes of hitherto unknown exact analytical time-dependent solutions in Maxwell electrodynamics of continuous media, discerned in principle from traditional models of sinusoidal waves, form the fundamentals of this novel field. These fundamentals demonstrate the strong dependence of reflection, refraction and absorption of transient waveforms on their shapes and time scales as well as on the natural time scales determined by the dispersion and conductivity of the material. Exactly solvable time-domain models of these phenomena are put forward in the non-Fourier electrodynamics discussed. Keeping in mind the aforesaid problems, one can outline the main goals of this book:

(1) to develop the physical foundations of time-domain electrodynamics for analysis of fast nonstationary processes in the electrodynamics of continuous media;

(2) to emphasize the importance of controlled shaping of transient pulses for optimization of pulse processes in nonstationary electrodynamics;

(3) to inculcate the concept of nonseparable fields in the physics of fast EM processes in media with plasma-like and waveguide-like dispersion.

This book may be reckoned as one of the first attempts at a self-contained introduction to a new field of physics forming a frontier between the optics of single waveforms and the theory of nonstationary propagation of an EM field in dispersive and conducting materials. Such waveforms of picosecond and subfemtosecond duration, produced by ultrafast photoconductive switching systems and optical compressors, are finding steadily increasing application. The following key concepts, inspired by consideration of the above-mentioned goals, run throughout this book.

1. Development of explicit exactly solvable models in non-Fourier electrodynamics. This novel topic does not use such traditional concepts of Fourier approach as frequency, wavelength, phase of oscillations, refractive index, phase velocity, etc. In contrast, the use of new nonseparable solutions of Maxwell equations and continuity conditions proves to be sufficient for direct time-domain consideration of the problems of transient electrodynamics in dispersive, guiding and lossy media.
2. The paramount role of the shapes of single-cycle waveforms, given their time scales, in the process of waveform reflection from dispersive and conducting interfaces. Use of these phenomena provides possibilities for real-time selection of discrete waveforms instead of traditional frequency filtration in systems for ultrashort signal processing.
3. The sensitivity of dynamics of nonstationary transmitted fields, excited by single waveforms in dispersive medium, to the time scales of the waveform and this medium. Such sensitivity is important for providing different rates of deformation of electric and magnetic components of transmitted fields.
4. Shape-dependent weakening of absorption of ultrashort EM transients in lossy media. These phenomena demonstrate the potentialities of optimization of pulse shapes for better penetration of conducting materials.
5. Simplification of the mathematical background of nonstationary electrodynamics due to the systematic use of new nonseparable field solutions of the Klein–Gordon and telegraph equations. Replacement of integration and differentiation of these fields by summation and subtraction of the finite variables, performed exactly because of the analytical properties of these fields, results in a drastic reduction of the relevant computer-memory requirements.

These concepts resemble some topics of wavelet physics, which is now entering a new era (Tilton 1986; Kaiser 1994). Use of wavelets for transient analysis is known to compete with sinusoidal waves in all-optical systems for signal processing. However, this book focuses mainly on the possibilities of using transients as carriers of information and energy in continuous media.

The restrictions of the age-old monopoly of sinusoidal waves in the analysis of time-dependent EM phenomena, as a result of the development of pulse electrodynamics, are very instructive. The nonstationary (time-domain) and quasi-stationary (frequency-domain) approaches may be considered as two branches of Maxwell electrodynamics. These new results show that despite the general interest of nonlinear phenomena, linear electrodynamics is still not exhausted.

This book consists of three parts. Chapter 1 is devoted to more or less traditional approaches to time-domain pulse electrodynamics (inverse Fourier transform, incomplete Fourier transform, finite-difference time-domain methods). In contrast, the non-Fourier electrodynamics of dielectric, conducting and guiding media, based on new explicit nonstationary solutions of the Maxwell equations, is advanced in Chapter 2. The exactly solvable model of reflection and refraction of arbitrarily shaped waveforms, based on these solutions, is presented in Chapter 3.

Interactions of ultrashort signals with continuous media are considered in Chapter 1 in the frequency domain by means of the concept of broadband pulses. A simple analytical model of slowly varying frequencies and spectral amplitudes of such pulses, typical of nonlinear optics, illustrates the mixed (spectral–temporal) description of the gradual evolution of a quasi-harmonic pulse in a dispersive medium. The examples of time-domain description of broadband pulse phenomena via the inverse FT of steady-state transfer functions as well as the use of the finite-difference time-domain approach reveal some merits and shortcomings of such computational electromagnetics.

To avoid these problems and develop the analytical time-domain electrodynamics of materials, widely used in optoelectronics and radiophysics, the Maxwell equations for Debye media, waveguide systems, and transmission lines are reduced in Chapter 2 to the normalized Klein–Gordon equation. Wide classes of exact nonsinusoidal solutions of the Klein–Gordon and telegraph equations, obtained beyond the scope of Fourier transforms, are presented. These nonseparable solutions, which are distinguished from a traditional product of time-dependent and coordinate-dependent parts, form the theoretical basis of nonstationary electrodynamics, describing the rapid distortions of EM fields in these media. Some salient features of these processes—the difference in propagation of electric and magnetic components of EM fields, pulse excitement of cutoff frequencies at the wings of the envelopes, vanishing of collisionless dissipation—are visualized in Chapter 2 as a result of these simple presentations of nonsinusoidal fields.

The excitation of time-dependent fields by arbitrarily shaped waveforms, represented in terms of a series of Hermitian or Laguerre pulses, is used in Chapter 3 for analytical determination of reflection and refraction coefficients for these waveforms. Unlike the analysis of these coefficients by means of complicated integration in the frequency domain, the use of this exactly solvable time-domain model involves standard algebraic procedures. The dependence of these coefficients on the time scales of the waveform and the medium and their sensitivity to the shape of the waveform, the time scales being given, is described by explicit formulae. The reflection–refraction properties of Hermitian and Laguerre signals are shown to be important for processing of ultrashort waveforms in real time.

Nonstationary electrodynamics is now taking its first steps. Many questions, especially regarding 2D and 3D problems, still remain unanswered. However, work is in progress, and nonstationary electrodynamics has a good chance of soon taking a place in science.

This book may be of interest to different classes of reader.

For *scientists* interested in basic physics, this book can provide a new tool for investigation of nonstationary pulse optical processes including time-dependent diffraction, reflection and filtration, governed by the temporal envelope of the localized field. The investigation of such pulse-shape-dependent phenomena gives rise to reconsideration of numerous wave problems treated traditionally by means of Fourier expansions. The development of non-Fourier optics of strongly dispersive media shows that even the investigation of simple power-independent properties of linear dispersive systems leads to novel and useful results.

For *designers* and *users* of pulse optical systems, this book for the first time in a detailed systematic manner offers in-depth discussions on physical aspects of the design of these systems. The controlled transformation of ultrashort waveforms is illustrated by a series of examples. Some important never-before-published principles of development of all-optical transient devices are considered.

For *lecturers* and *students*, the arrangement of information is suitable for didactic purposes and self-instruction. The reasons for using either temporal or spectral descriptions, either Fourier or non-Fourier approaches for analysis of the problems discussed are considered by way of physically meaningful examples, Comparison of the two approaches is included in most sections. Detailed explanations of mathematical methods and numerous illustrations are given to simplify the presentation and promote the formation of valid 'pulse physics intuition'.

During the last two years many people have contributed to the author's understanding of this field. I thank Professors R. Albanese, F. Bunkin, C. Holland, G. Kaiser, S. Mallat, A. Masalov, A. Rukchadze, H. Szu and L. Stenflo for illuminating discussions. It is my pleasure to thank Prof. T. Wilson and Dr R. Lawrence for their moral support throughout this work. The fruitful stimulating criticism of Prof. S. Ritov is highly appreciated. Once again I thank my family for patience and encouragement.

Moscow A. B. S.
November 1995

CONTENTS

1

FUNDAMENTALS OF OPTICS OF BROADBAND HARMONIC PULSES

1.1 Introduction: time-domain and frequency-domain optics of nonstationary processes

This chapter is devoted to time-dependent phenomena in interactions of short pulses of quasi-monochromatic electromagnetic radiation, containing only a few oscillations of wave fields, with dispersive and lossy media. These phenomena are now discovered in a multitude of pulse reflection, scattering and absorption processes. The traditional steady-state values of frequency-dependent parameters characterizing these processes—the reflection coefficients, scattering cross sections and absorption decrements—are determined by the properties of materials irradiated by harmonic CW radiation. In contrast, the relevant characteristics of nonstationary processes may depend on the duration and shape of incident waveform, the material and the geometry of the experiment being the same. To consider these time-dependent effects in the framework of frequency-domain approach, one should use a Fourier transform for the incident field $E_1(t)$ and the subsequent inverse Fourier transform for re-emitted spectral component $E_2(\omega)$

$$\left.\begin{aligned} E_1(\omega) &= \int_{-\infty}^{\infty} E_1(t)\, e^{i\omega t}\, dt \\ E_2(t) &= \frac{1}{2\pi}\int_{-\infty}^{\infty} E_2(\omega)\, e^{-i\omega t}\, d\omega. \end{aligned}\right\} \tag{1.1}$$

The correlation between the spectral components $E_1(\omega)$ and $E_2(\omega)$ is obtained from consideration of the pulse effect in the frequency domain. Unlike the long trains of CW radiation with one carrier frequency ω, solitary waveforms with finite characteristic duration T are described by nonzero spectral bandwidth $\Delta\omega \propto T^{-1}$. The effect of pulsewidth on bandwidth is known to be evaluated by the ratio $\Delta\omega\,\omega^{-1}$:

$$\frac{\Delta\omega}{\omega} = \frac{k}{\omega T}\,. \tag{1.2}$$

The dimensionless coefficient k—the so-called spectral-time product—depends on the waveform shape, the duration T being fixed. The fractional bandwidth $\Delta\omega\,\omega^{-1}$ reaches a value of roughly unity for the single-cycle pulse $\omega T \approx 2\pi$.

It should be mentioned that a steady-state flux of polychromatic radiation may possess a broadband spectrum too. Hence the results obtained for these fluxes can be used for qualitative interpretation of some trends in interactions of short broadband pulses with optical systems in the frequency domain (Wolf 1986). However, this spectral presentation is nonlocal, contrary to the explicit nonstationary picture of such interactions in the time domain. This time-dependent picture would be reinstated as a result of the inverse Fourier-transform of steady-state results, eqn (1.1). Such a procedure, used for example in time-resolved spectroscopy (Dodd and Series 1979), is accomplished as a rule by numerical integration with respect to frequency. The derivation of explicit information from this presentation is complicated when the incident pulse is narrowing and, correspondingly, its bandwidth is broadening (Howell and Uberall 1990).

To visualize the physically meaningful trends in the interactions of short electromagnetic pulses with dispersive and absorbing media, a straightforward consideration of these phenomena in the time domain would be needed. Instructive examples of such trends are provided by time-resolved scattering and diffraction effects (Chesnoy and Mokhtari 1987). The advantages of direct integration of the Maxwell equations in the time domain for a series of reflection and transmission problems are illustrated mainly by numerical methods (Sun and Rusch 1994); the new analytical approach to these problems is developed in Chapter 3.

To remove some problems in rigorous frequency-domain as well as time-domain approaches, the 'mixed' description of quasi-stationary problems may become convenient. Thus, operating with slowly varying amplitude profiles of wideband solitary waveforms or time-dependent frequencies of modulated pulses, one can combine the description of pulse shaping in the time domain with frequency-domain dispersion effects. Since short pulses are generated mainly by high-power optical systems, this 'mixed' approach is widely used in nonlinear optics (Stegeman *et al.* 1988).

Ultrashort waveforms with picosecond and subpicosecond time scales will be discussed in this Chapter. The quasiharmonic laser pulses in these ranges are usually produced by so-called mode-locking. Lasers are known to allow gain at many different mode frequencies, each separated by a frequency spacing $c(2L)^{-1}$, so that an integral number of half-wavelengths of each of these mode frequencies are able to fit between laser mirrors separated by a distance L. To provide mode-locking, one should use for example laser gain modulation, generating the sidebands of each mode; the sidebands coincide in frequency with next-neighbour mode frequencies and share the energy coherently among the modes until a common phase is established. These correlated modes will interfere to produce a regular pulse train. The duration of each of the pulses depends inversely on the number of oscillating modes locked in-phase. Thus, to obtain 100 fs pulses we require, according to eqn (1.2), a bandwidth $\Delta\omega = 2\pi \times 10^{13}$ rad s^{-1}. If the cavity length is $L = 1$ m, the mode spacing $c(2L)^{-1}$ is ~10^9 rad s^{-1} and we need to lock in-phase 7×10^4 modes. Obviously a large gain bandwidth is necessary for ultrashort pulse generation, and this is why dye lasers with a huge gain across much of the optical wavelength region are employed for these purposes.

The powerful laser pulses may be compressed owing to nonlinear propagation along the optical fibre and subsequent passage through the pair of diffraction gratings. This pulse compression method has been borrowed from chirp radar technique. The phase self-modulation of the pulse, travelling along the fibre with power-dependent refractive index, provides the development of frequency modulation $\Delta\omega$ from lower to higher frequencies from the front edge to the trailing edge of the pulse. The geometry of the grating pair is arranged so that the 'red' edge of the pulse has further to travel owing to the grating's dispersion, whereas the 'blue' edge has a shorter optical path length and catches up (Grischkowsky and Balant 1982); the output pulsewidth is compressed to ~$(\Delta\omega)$.

Considering ultrafast electromagnetic signals, one should emphasize that the methods of shaping such signals in the optical and radio ranges are essentially different. Unlike the aforesaid compressed laser pulses, the recent successes in formation of picosecond microwave transients are based on the optoelectronically switched resonator (Paulus *et al.* 1986) or on Cerenkov generation of an electromagnetic shock wave from an electric pulse propagating on a coplanar transmission line (Fattinger and Grischkowsky 1989). Nevertheless, despite these distinctions, the physical fundamentals of reflection, scattering and absorption of such wideband pulses by dispersive and lossy media in both the optical and microwave ranges contain several similar features. To develop an insight into time-domain and frequency-domain optics of ultrafast EM wave processes and their use for design of all-optical systems, this Chapter is organized as follows.

Since the nonstationary wave processes in the fields of short solitary waveforms will be shown to depend effectively on their shapes and time scales, the control of these characteristics is a vital factor in pulse all-optical operations. To optimize these operations, the waveforms with extremal spectral properties are revealed in Section 1.2. Shape-dependent spectral bandwidths are determined for both sinusoidal and nonsinusoidal field pulses.

The controlled shaping of broadband pulses, including the formation of their time-dependent spectra in nonlinear fibre systems, is considered in Section 1.3. The sensitivity of this process to initial pulse profiles is shown to provide the development of frequency modulation both near the envelope maximum and at the pulse wings. The dynamics of such processes is analysed by means of exact analytical nonstationary solutions of equations of nonlinear geometric optics. The conditions of shaping of spike-like pulses as well as steepening of pulse wings are outlined.

Fast generation of pulses due to self-lamination of powerful radiation fluxes, stimulated by modulational instability of trapped waves in fibre systems, is considered in Section 1.4. The transformations of such fluxes to trains of similar pulses depending on initial frequency modulation are illustrated. The enhancement of these processes in systems of periodic structure is examined.

In contrast to Sections 1.3 and 1.4, devoted fo the principles of formation of short modulated wave pulses, some processes of reflection and scattering of such pulses are considered in Sections 1.5 and 1.6. Consideration is given to the peculiarities of these processes, associated with nonzero spectral bandwidths and finite time scales of the waveforms interacting with dispersive systems. The diffractive variations of the spectra of polychromatic wave beams are shown to be accumulated in the course of propagation in dispersive media subject to the intensity profile of the beam. The smoothing of the diffraction pattern of broadband pulses as compared with the standard monochromatic picture is demonstrated (Section 1.5). Reshaping of transient fields reflected from resonant media and scattered on resonant targets is discussed in Section 1.6. The lengthening of the reflected pulse due to excitement of eigenfrequencies of the scattering system, forming the 'ringing tail' of this pulse, is illustrated for both penetrable and impenetrable fields. The different mechanisms of time-resolved reflectivity owing to excitonic resonances and vibrational transitions are also mentioned in Section 1.6.

In contrast to the broadband pulse effects considered above in the frequency domain, the shape-dependent absorption of short waveforms is analysed in Section 1.7 directly in the time domain. This effect is demonstrated for two- and three-level quantum systems, and the possibilities of using absorption of these waveforms for time-resolved spectroscopy of vibrational states as well as for excitement of some forbidden transitions are considered.

The current state of the art in time-domain and frequency-domain optics of short pulses interacting with dispersive and lossy media is based mainly on computer simulations. These approaches are used in Section 1.8 for the treatment of two instructive problems: the direct numerical integration of finite-difference equations replacing the Maxwell equations, and analysis of short-pulse propagation in plasma by means of Fourier expansions. The computational and principal problems of both approaches are outlined; this analysis gives rise to the development of non-Fourier optics advanced in the subsequent chapters.

1.2 Spectral properties of modulated sinusoidal waves and nonsinusoidal waveforms

The shaping of electromagnetic pulses is accompanied by tuning of their spectral bandwidths. The steepening of pulse fronts promotes broadening of spectral bandwidth; their flattening can give rise to narrowing of the spectrum. Wave trains with narrow bandwidth may be of use for numerous interactions of EM fields with resonant systems; the use of broadband waveforms is important for communication and remote sensing applications.

Freely propagating picosecond pulses can be generated by means of a number of pulse-forming devices based on the charged line concept. One could mention for example the radiation of extremely short-lived electric dipoles produced by shorting of charged, micrometre-sized coplanar transmission lines with ultrashort laser pulses (Fattinger and Grischkowsky 1988). Other methods of using transmission lines are associated with the formation of picosecond electrical pulses due to a series of capacitance discontinuities, embedded in such lines (Frankel *et al.* 1989) or, in contrast, the continuous shaping of such pulses due to dispersion of a microstrip line (Whitaker *et al.* 1987).

To optimize the waveforms $E = E_0 F(t)$ with respect to the band-limited communication system or to the breadth of absorption line, one should use the Fourier transform (FT) of this waveform $E(\omega) = E_0 F_\omega$ eqn (1.1). The energy $\mathcal{E}(\Delta\omega)$ contained in some frequency range $(-\Delta\omega, \Delta\omega)$ around the carrier frequency may be expressed via the spectral density of the energy $|E_0|^2 |F_\omega|^2$

$$\mathcal{E}(\Delta\omega) = |E_0|^2 W(\Delta\omega), \tag{1.3}$$

$$W(\Delta\omega) = \frac{1}{2\pi} \int_{-\Delta\omega}^{\Delta\omega} |F_\omega|^2 \, d\omega; \tag{1.4}$$

the normalized total energy of the pulse W is given by the value $W(\infty)$. The spectral bandwidth of the envelope $F(t)$ is defined usually as the range of frequencies $(-\Delta\omega_0, \Delta\omega_0)$ containing 99% of the total pulse energy:

$$W(\Delta\omega_0) = 0.99 W(\infty). \tag{1.5}$$

The value $\Delta\omega_0$ may be found by the solution of eqn (1.5).

Some properties of pulse envelopes having an essential influence on their spectral bandwidths, such as the rise time, are considered below by means of traditional models of amplitude-modulated and phase-modulated sinusoidal waves, widely used in optics and radio physics (Subsection 1.2.1.);

the narrowest pulses in both frequency and time domains are also analysed. Some useful presentations of nonsinusoidal electromagnetic waveforms are described in Subsection 1.2.2.

1.2.1 Spectra of amplitude-modulated and phase-modulated harmonic oscillations

Considering the concept of modulated waves, important for transfer of energy and information, one could represent the wave envelope F in a dimensionless form:

$$F = f \exp(-\mathrm{i}\omega t + \mathrm{i}\varphi). \tag{1.6}$$

The case of constant values of f, ω and φ relates to the model of a long monochromatic wave train. This model, described by the spectrum

$$F_\omega = f\delta(\omega), \tag{1.7}$$

where $\delta(\omega)$ is the delta function, corresponds formally to a zero-bandwidth model of an unmodulated harmonic wave field, forming the basis of many topics in traditional optics. However, each temporal variation of amplitude f, phase φ or frequency ω in eqn (1.6) may be characterized by a set of harmonic oscillations. Such waves with nonzero bandwidths are radiated by real sources. Deferring consideration of frequency modulation to Section 1.3; let us analyse initially the dependence of the spectral properties of quasi-monochromatic wave fields on their amplitude–phase modulation.

We may discuss first a square wave pulse with rapid rise time, represented by

$$F = \begin{cases} \cos(\omega_0 t + \varphi), & -T \leqslant t \leqslant T \\ 0, & t^2 > T^2 \end{cases} \tag{1.8}$$

This model is useful for many applications in communication systems and high-speed signal processing. Such square wave pulses are especially promising for optical circuits, because they are not susceptible to capacitive effects, which are known to limit the rise time of electronic signals in transmission lines (Pang *et al.* 1991). Moreover, the determination of the spectral bandwidth of the pulse, eqn (1.8), provides a simple example of such analysis. Substituting eqn (1.8) in eqn (1.1) we obtain the FT of the square wave:

$$F_\omega = \frac{\sin[(\omega - \omega_0)T]}{\omega - \omega_0}(\cos\varphi + \sin\varphi) + \frac{\sin[(\omega + \omega_0)T]}{\omega + \omega_0}(\cos\varphi - \sin\varphi) \tag{1.9}$$

The function F_ω, eqn (1.9), shown in Fig. 1.1, illustrates the important role of phase modulation for short pulses, containing for example only one

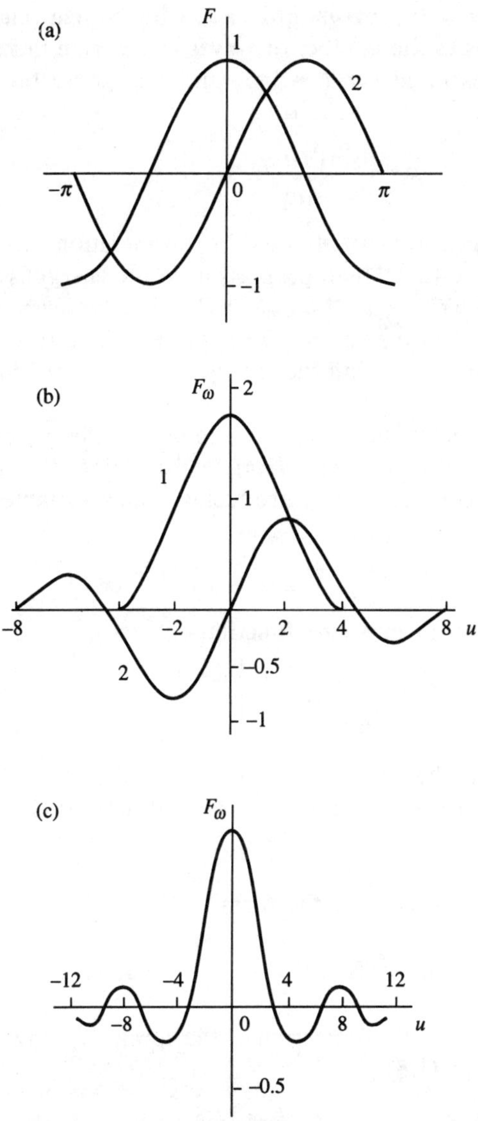

Fig. 1.1 Fourier transform F_ω of a wave pulse F (eqn 1.8): (a) the temporal envelopes of single-cycle pulses ($\delta = \omega_0 T = \pi$) for $\varphi = 0$ (curve 1) and $\varphi = -\pi/2$ (curve 2); (b) Fourier spectra of single-cycle pulses; curves 1 and 2 relate to envelopes 1 and 2 in (a); (c) graph of F_ω for a long wave pulse ($\omega T \gg \pi$).

oscillation ($\omega_0 T = 2\pi$) of an EM field (single-cycle pulses). This phase effect diminishes as the number of waves in the train increases. When this number becomes large ($\omega_0 T \gg 2\pi$), one can ignore the contribution of the second term in eqn (1.9) to the value of F_ω and express F_ω in the form

$$F_\omega = \frac{\sin[T\Delta\omega]}{\Delta\omega}, \quad \Delta\omega = \omega - \omega_0. \tag{1.10}$$

This Fourier transform, widely used in consideration of long wave trains, is shown in Fig. 1.1b. A high proportion of the energy is contained in the range $\omega = \omega_0 - \pi T^{-1}$ and $\omega = \omega_0 + \pi T^{-1}$, i.e. between two minima surrounding the central maximum. One can see, using this example, that in the case of a long wave train the energy is concentrated in a narrow range of $\Delta\omega$.

To find the spectral bandwidth $\Delta\omega_0$ eqn (1.5), let us calculate the normalized energy $W(\Delta\omega)$ eqn (1.4), substituting the function F_ω of eqn (1.10) into eqn (1.4) and introducing a new variable $x = T\Delta\omega$:

$$W(\Delta\omega) = \frac{T}{2\pi}\left[\mathrm{Si}(2x) - \frac{\sin^2 x}{x}\right]. \tag{1.11}$$

Here $\mathrm{Si}(x)$ is the integral sine function

$$\mathrm{Si}(x) = \int_0^x \frac{\sin y}{y}\,\mathrm{d}y.$$

Keeping in mind the value $W(\infty)$ from eqn (1.11), and designating the normalized spectral bandwidth as a spectral-time product

$$k = T\Delta\omega_0, \tag{1.12}$$

one can rewrite eqn (1.5) in the form

$$\frac{2}{\pi}\left[\mathrm{Si}(2k) - \frac{\sin^2 k}{k}\right] = 0.99. \tag{1.13}$$

The solution of eqn (1.13) determines the spectral bandwidth of a square-shaped pulse, eqn (1.8):

$$k = 32.8. \tag{1.14}$$

This result corresponds to eqn (1.2) with the value of parameter $k = 32.8$. Proceeding in a similar fashion, one can find the value k for the triangular pulse, $k = 10.2$. The decrease in k, as compared with eqn (1.14), is connected with the slope of the sides of the triangular profile.

It is necessary to point out that FT (1.10), widely used (e.g. in the theory of diffraction), is a limiting case of the more general result, eqn (1.9), for an arbitrary number of field oscillations in a train. Substituting eqn (1.9)

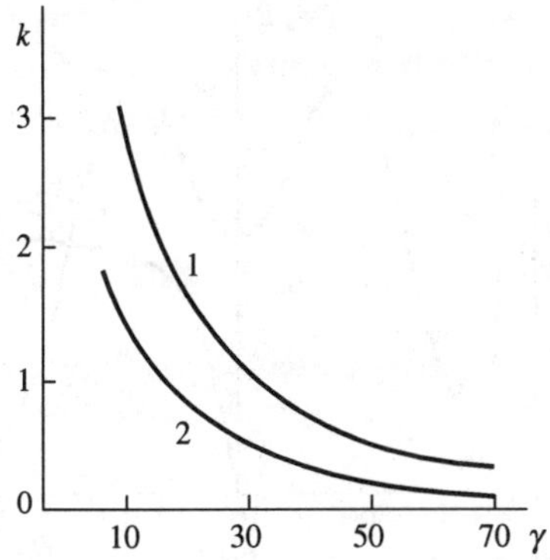

Fig. 1.2 Spectral bandwidth k (eqn 1.12) of a phase-modulated square-wave pulse vs. number of cycles in a pulse γ. Curves 1 and 2 relate to $\varphi = 0$ and $\varphi = -\pi/2$ respectively.

into eqn (1.4), one can find the spectral bandwidths of short waveforms containing only one or a few oscillations of an EM field. These results, shown in Fig. 1.2, illustrate the influence of phase modulation on the spectral properties of single-cycle waveforms.

The high bandwidth of a square-shaped pulse ($k \gg 1$) and short rise time (tending formally to zero) show the advantages of using these pulses in all-optical information systems. To visualize the influence of increasing rise time and growth of the slope of the fronts of the pulse it is expedient to consider a model of a smooth amplitude-modulated phase-dependent wave train:

$$F = \exp\left(-\frac{t^2}{2T^2}\right) \cos(\omega_0 t + \varphi). \tag{1.15}$$

The Fourier transform of this cos-Gaussian function is

$$\left.\begin{aligned} F_\omega = T\sqrt{2\pi}\, \exp\left(-\frac{u^2+\delta^2}{2}\right) [\cos\varphi\, \mathrm{ch}(\delta u) - \mathrm{i} \sin\varphi\, \mathrm{sh}(\delta u)], \\ u = \omega T; \quad \delta = \omega_0 T. \end{aligned}\right\} \tag{1.16}$$

In contrast to the real FT, eqn (1.9), the function F_ω, eqn (1.16), being complex, may be represented by means of its amplitude A_ω and phase χ_ω:

$$\left.\begin{aligned} F_\omega = A_\omega \exp(-\mathrm{i}\chi_\omega), \\ A_\omega = T\sqrt{2\pi}\, \exp\left(-\frac{u^2+\delta^2}{2}\right) \sqrt{[\mathrm{ch}^2(\delta u) - \sin^2\varphi]}, \\ \mathrm{tg}\,\chi_\omega = \mathrm{tg}\,\varphi\, \mathrm{th}(\delta u). \end{aligned}\right\} \tag{1.17}$$

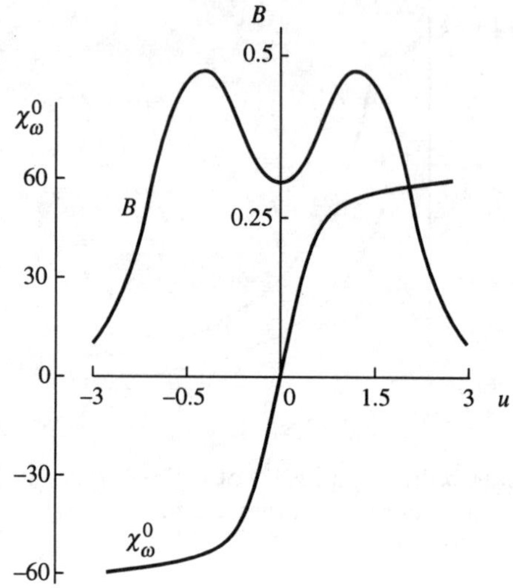

Fig. 1.3 Fourier spectrum of an amplitude- and phase-modulated pulse (eqn 1.15); normalized amplitude $B = A_\omega T^{-1}(2\pi)^{-1/2}$ and phase χ^0_ω (eqn 1.18) are plotted against the spectral parameter $u = \omega T$; $\varphi = 60°$, $\delta = \omega_0 T = 1$.

The amplitude A_ω and phase χ_ω of the function F_ω, shown in Fig. 1.3, are respectively even and odd functions of the dimensionless spectral parameter $u = \omega t$. It is worth while to emphasize that the phase χ_ω arises due to initial phase modulation in eqn (1.15). The dependence of amplitude A_ω on the phase φ, eqn (1.17), becomes negligible when $\delta u \geqslant 1.5$.

To reveal the influence of amplitude–phase modulation on the spectral bandwidth, one should calculate the energy function W, eqn (1.4), by means of FT (1.16). Using the error function $\mathrm{erf}(x)$:

$$\mathrm{erf}(x) = \frac{2}{\sqrt{\pi}} \int_0^x \mathrm{e}^{-z^2}\,\mathrm{d}z, \quad \mathrm{erf}(-x) = -\mathrm{erf}(x),$$

one can reduce eqn (1.5) to the form

$$1 - 0.5[\mathrm{erf}(u+\delta) + \mathrm{erf}(u-\delta)] + \exp(-\delta^2)[1 - \mathrm{erf}(\mathrm{u})]\cos^2\varphi$$
$$= 10^{-2}[1 + \exp(-\delta^2)\cos^2\varphi]. \qquad (1.18)$$

The solutions of eqn (1.18), describing the dependence of dimensionless spectral bandwidth k upon the parameter $\delta = \omega_0 T$ for the extremal values of phase $\varphi = 0$ and $\varphi = -\pi/2$, are shown in Fig. 1.4.

Models of pulses assuming equality of rise time and fall time have been considered so far. However, there is a multitude of real electromagnetic

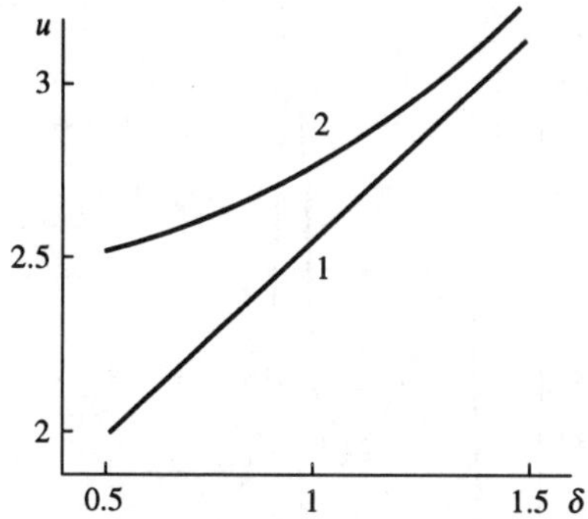

Fig.1.4 Dependence of dimensionless spectral bandwidth k (eqn 1.12) of amplitude- and phase-modulated pulses (eqn 1.15) on the spectral parameter $\delta = \omega_0 T$. Curves 1 and 2 relate to phases $\varphi = 0$ and $\varphi = -\pi/2$ respectively.

pulses that do not possess such symmetry. Thus the photoconductive switch, operating in many all-optical systems, is characterized by two different time scales. When this switch is excited by picosecond laser pulses, an electric pulse with approximately the same rise time as that of the optical pulse will be generated; the fall time of the electric pulse is determined by the carrier lifetime, which may extend to a few hundred picoseconds for certain photoconductive materials. As a result, the electrical pulses generated are of asymmetric shapes which can be described approximately by a Gaussian rising edge followed by an exponentially decaying tail (Sano *et al.* 1989). These so-called single-sided exponential pulses (Fig. 1.5) are widely considered in picosecond optics (Qian and Yamashita 1991) and in the physics of waveforms irradiated by ultrawide bandwidth radar (Ma and Ciric 1992).

Two spectral properties of amplitude-modulated (AM) and phase-modulated (PM) pulses, discussed above, should be outlined:

(1) an increase in the rise time, the characteristic time scale of the pulse being fixed, provides spectral narrowing;

(2) the longer an amplitude-modulated wave train, the less sensitive proves to be its spectrum to phase perturbations of this train.

These general trends reveal ways to broaden or to narrow the spectral bandwidths $\Delta\omega_0$ of pulses of a given duration by shaping of their envelopes. In the case of short pulses, such possibilities can be improved by phase modulation. However, each optical system is characterized by certain inherent values $\Delta\omega_0$, and to choose the optical pulses for this system, some extremal properties of the pulses may become useful.

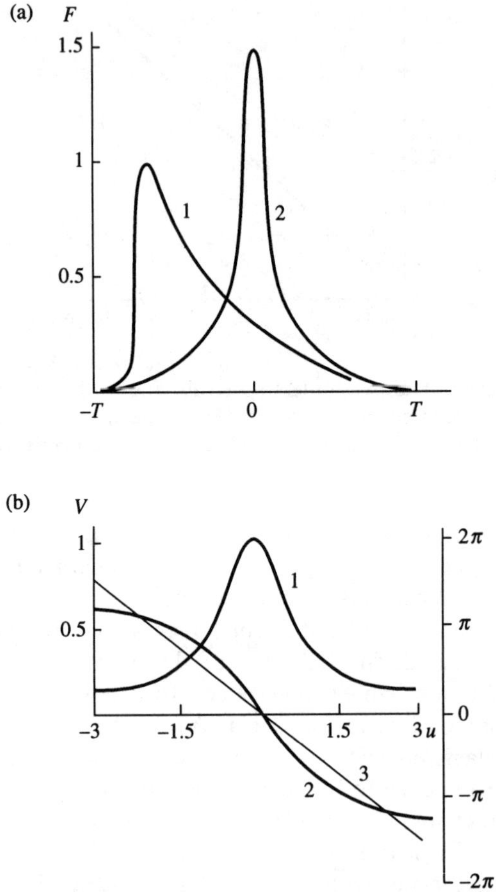

Fig. 1.5 Spectral properties of the single-sided exponential pulse shown in (a), curve 1. (b) Normalized spectral envelope $V = F_\omega(0) \cdot F_\omega^{-1}(\omega)$ (curve 1) and phase (curve 2) of a single-sided exponential pulse plotted vs. the parameter $u = \omega T$; curve 3 is the phase of limited Fourier transform pulse with spectral envelope coinciding with curve 1. The symmetrization and narrowing of the temporal envelope of the Fourier transform limited pulse is shown by curve 2 in (a).

In designing power-carrying optical systems, it is worth while to find the shape of a pulse $F(t)$, its time scale $2T$ being given, containing the maximal fraction of its energy $\mathcal{E}$ in the given frequency range $(-\Delta\omega_0, \Delta\omega_0)$. Considering this pulse to be 'ideal' for such a system, let us write its spectral density $|F_\omega|^2$ by means of eqn (1.1):

$$|F_\omega|^2 = \int_{-T}^{T}\int_{-T}^{T} F(t)F(s)\,e^{i\omega(t-s)}\,dt\,ds. \tag{1.19}$$

Substituting $|F_\omega|^2$ from eqn (1.20) into eqn (1.4) and integrating with respect to ω, one can find the function $W(\Delta\omega_0)$:

$$W(\Delta\omega_0) = \frac{1}{\pi}\int_{-T}^{T}\int_{-T}^{T} \frac{\sin\omega_0(t-s)}{t-s}F(t)F(s)\,dt\,ds. \tag{1.20}$$

Our goal is to find the envelope $F(t)$ providing the maximal value of $W(\Delta\omega_0)$, eqn (1.20), and satisfying the normalization condition

$$\int_{-T}^{T} F^2(t)\,dt = 1. \tag{1.21}$$

Using the symmetry of the kernel of eqn (1.20),

$$\kappa(t, s) = \frac{1}{\pi}\frac{\sin\omega(t-s)}{t-s}, \tag{1.22}$$

one can reduce eqn (1.20) (Courant and Hilbert 1953) to the Fredholm integral equation

$$F(t) = \frac{\mu}{\pi}\int_{-T}^{T}\frac{\sin\omega(t-s)}{t-s}F(s)\,ds. \tag{1.23}$$

Here the unknown envelope $F(t)$ is the first fundamental solution of eqn (1.24), associated with the value

$$\mu = \frac{1}{W(\Delta\omega_0)}. \tag{1.24}$$

Choosing, according to eqn (1.5) and normalization condition (1.21), the value $W(\Delta\omega_0) = 0.99$, one can solve eqn (1.24) numerically and find the 'ideal' pulse $F_{id}(t)$ and its spectral bandwidth k_{id}. This extremal pulse, determined in the interval $(-T, T)$ and characterized by the least value $k_{id} = 3.6$, is shown in Fig. 1.6. Since the explicit analytical formula for $F_{id}(t)$ is unknown, it may be described approximately in the same interval $(-T, T)$ by means of the cosine function (Fig. 1.6)

$$F_{id} \approx \cos\left(\frac{\pi}{2}\frac{t}{T}\right). \tag{1.25}$$

The bandwidth of this envelope, $k = 4$, is somewhat greater than $k_{id} = 3.6$. The Gaussian envelope with the value $k = 20$ is given for comparison in Fig. 1.6 also.

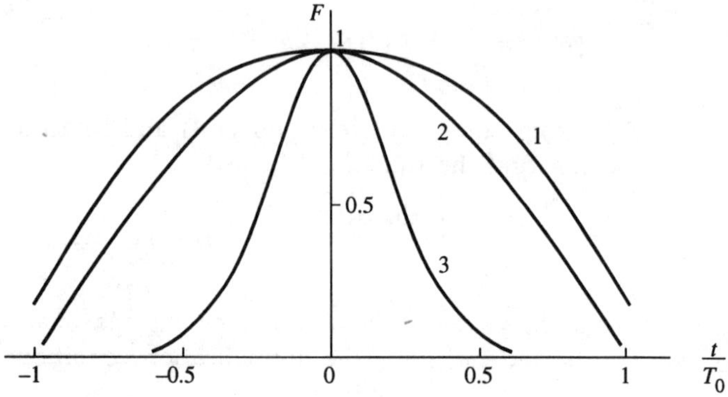

Fig. 1.6 Spectral bandwidths k of symmetrical pulses with the same time scale but different envelopes. Curve 1 relates to the 'ideal' pulse, containing 99% of its energy in the given spectral range ($k = 3.6$); curve 2 (eqn 1.25) and Gaussian envelope 3 relate to $k = 4$ and $k = 20$ respectively.

Another extremal waveform is associated with so-called Fourier transform limited pulse, which is the narrowest pulse possible for a given spectral bandwidth. Thus the propagation of an asymmetric single-sided pulse (Fig. 1.5a, curve 1) along the transmission line provides a continuous dispersive delay of phase φ; meanwhile the amplitude spectrum (Fig. 1.5b, curve 1) remains invariant, the nonlinearity and losses of the line being negligible. This effect results in a reshaping of the pulse; the moment when φ becomes a linear function of frequency (Fig. 1.5b, curve 3) relates to compression of the pulse to the smallest width in the time domain (Fig. 1.5a, curve 2). Unlike the nonlinear compression of pulses (Section 1.3), the mechanism discussed is power-independent.

These results indicate the possibilities of optical shaping of sine pulses: their bandwidths are shown to vary by a factor of 8–9 (from $k = 3.6$ for an 'ideal' pulse to $k = 32.8$ for a square-shaped one), their time scales $(-T, T)$ and peak amplitudes remaining the same. Pulse spectral bandwidth measurements are often associated with measurements of pulse duration. Relevant experiments have been based usually on optical interferometry and heterodyning devices. However, recent investigation of the dependence of the angular divergence of the second harmonic beam upon the spectral properties of pumping pulses has opened up new possibilities for determination of the time scales and spectral structures of these pulses (Kabelka and Mosolov 1993); using this effect, the spectral width, pulse duration, spectral-time product, chirp and its sign were shown to be measured simultaneously in the same device.

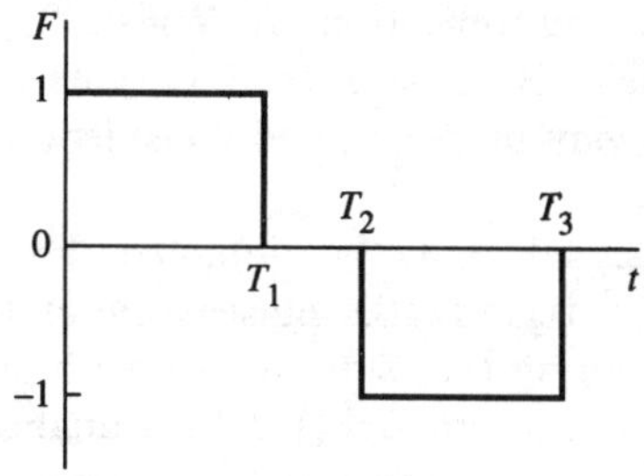

Fig. 1.7 Walsh signals.

1.2.2 Nonsinusoidal waveforms and Hermitian signals

The use of sequences of nonsinusoidal single-cycle waveforms is considered to be promising for processing short time-varying signals in real time. These waveforms are useful for signal multiplexing, pattern recognition and image enhancement. Unlike the amplitude-modulated trains of sinusoidal waves, the standardized models of nonsinusoidal envelopes prove to be especially efficient in analysis of pulses with time-dependent frequencies, which cannot be represented by any caxrier frequency. Although the dynamics of such pulses in the real media will be discussed in details in Chapter 3, some spectral properties of inharmonic waveforms are illustrated below.

The first examples of such pulses, widely used in radio physics and quantum mechanics, were associated with the Heaviside unit step function

$$H(t) = \begin{cases} 1, & t \geqslant 1, \\ 0, & t < 0, \end{cases} \tag{1.26}$$

and the Dirac delta function $\delta(t)$. These functions describe formally the discontinuities in distributions of electromagnetic fields. The combination of several Heaviside functions (1.26) forms the Walsh pulse (Walsh 1923):

$$F(t) = \begin{cases} 1, & 0 < t < T_1, \\ -1, & T_2 < t < T_2 + T_1 \end{cases} \tag{1.27}$$

and $F(t) = 0$ outside the intervals (1.28) (Fig. 1.7). In the limiting case $T_1 = T_2$, related to a jump in the envelope $F(t)$ (1.27) from the value $F = 1$ to the value $F = -1$, formula (1.27) describes the antisymmetric Haar function. Walsh signals are known to form a closed set of orthonormal functions, so they may be used for representation of the wide classes of nonsinusoidal envelopes beyond the scope of Fourier expansions (Harmuth 1977). The advantage of Walsh function expansion over other orthogonal function expansions lies in the fact that the Walsh function assumes only the values ± 1; hence multiplication by Walsh functions involves only algebraic sign assignment. Because of the binary nature of these functions, the Walsh transforms may be performed considerably faster than Fourier ones (Blachman 1974).

Spectra of Walsh signals may be obtained directly from eqn (1.10) as a limiting case of a rectangular pulse without the carrier frequency ($\omega_0 = 0$). It is worth noting that side by side with advantages in processing of signals, attempts to use model (1.27) for analysis of electromagnetic processes in real media face some problems, connected with the discontinuities of envelope (1.27) in the time domain at the edges. In contrast, the spectrum of pulse (1.27) is continuous over the whole frequency domain ($-\infty < \omega < \infty$), as well as its derivative wlth respect to ω.

An opposite tendency in the shaping of localized pulses is exhibited by the complex envelope, determined formally over the whole time domain ($-\infty < t < \infty$) but concentrated locally around the point $t = 0$ (Newland 1993):

$$F = (4\pi i\tau)^{-1}[\exp(\pi i\tau) - \exp(2\pi i\tau)], \quad \tau = t/T, \tag{1.28}$$

where T defines some scaling time. The real and imaginary parts of eqn (1.28), represented by odd and even functions of τ respectively, are continuous, unlike eqn (1.27); their spectra have box-like forms, confined to the octave bands. For example, the imaginary (odd) part of eqn (1.28),

$$\mathrm{Im}\,\mathrm{F} = (2\pi\tau)^{-1}(\sin 4\pi\tau - \sin 2\pi\tau), \tag{1.29}$$

and its Fourier transform are shown in Fig. 1.8. Contrary to eqn (1.10), the derivatives of the spectral envelopes of both real and imaginary parts of F, eqn (1.28), with respect to ω have discontinuities at the edges $\omega = \pm 2\pi, \pm 4\pi, \ldots$

Such waveforms, localized in both the time and frequency domains, so-called wavelets, can be optically implemented to represent short transient sigpals (Sheng *et al.* 1992). The above-mentioned amplitude-modulated waves, eqn (1.15) can also be considered as wavelets (Martinet *et al.* 1987). Unlike these simple waveforms, a family of continuous but not very regular (neither symmetric nor antisymmetric) wavelets was built up by Daubechies (1988).

Unlike these wavelets, the pulses shaped by Hermitian functions $\Phi_m(\tau)$ provide a flexible way of representing some waveforms with ripidly varying frequencies. These functions are continuous; τ is the normalized time, eqn (1.28), and the number m is the number of real zeros of the function $\Phi_m(\tau)$ (Fig. 1.9):

$$\left.\begin{array}{ll}\Phi_0(\tau) = C_0\exp(-\tau^2/2), & \Phi_1(\tau) = 2C_1\exp(-\tau^2/2),\\ \Phi_2(\tau) = C_2(4\tau^2 - 2)\exp(-\tau^2/2), & \Phi_3(\tau) = C_3(8\tau^3 - 12\tau)\exp(-\tau^2/2).\end{array}\right\} \tag{1.30}$$

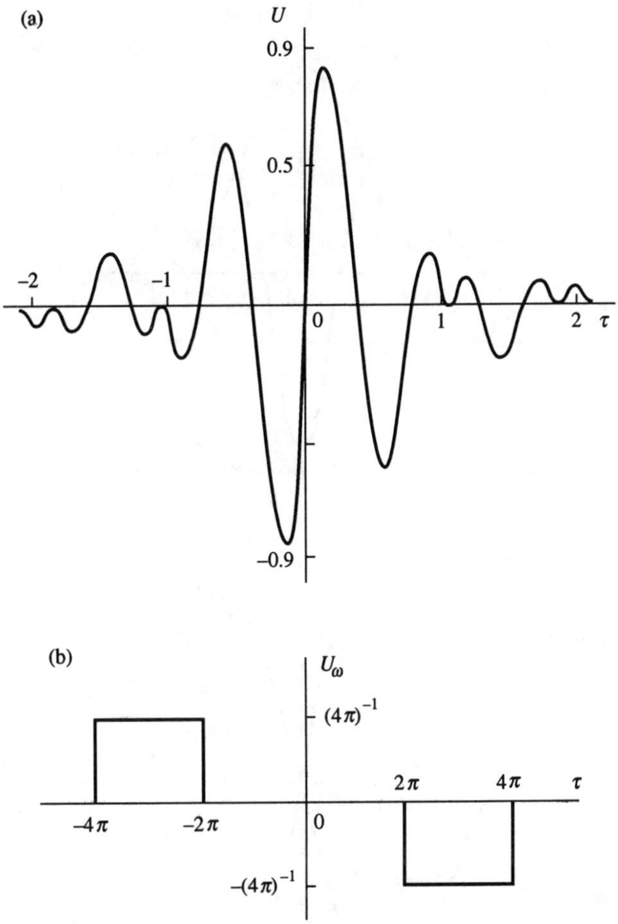

Fig. 1.8 (a) Envelope of the function $U = \operatorname{Im} F$ (eqn 1.30); (b) it's Fourier spectrum U_ω.

The Hermitian functions are known to form an orthonormal basis over the whole time domain $(-\infty < \tau < \infty)$, and the coefficients C_m are determined from the normalization condition

$$\left.\begin{aligned} &\int_{-\infty}^{\infty} \Phi_m(\tau)\Phi_n(\tau)\,\mathrm{d}\tau = \begin{cases} 1, & m = n, \\ 0, & m \neq n, \end{cases} \\ &C_0 = \frac{1}{\sqrt[4]{\pi}}, \quad C_m = \frac{C_0}{\sqrt{2^n m!}}. \end{aligned}\right\} \tag{1.31}$$

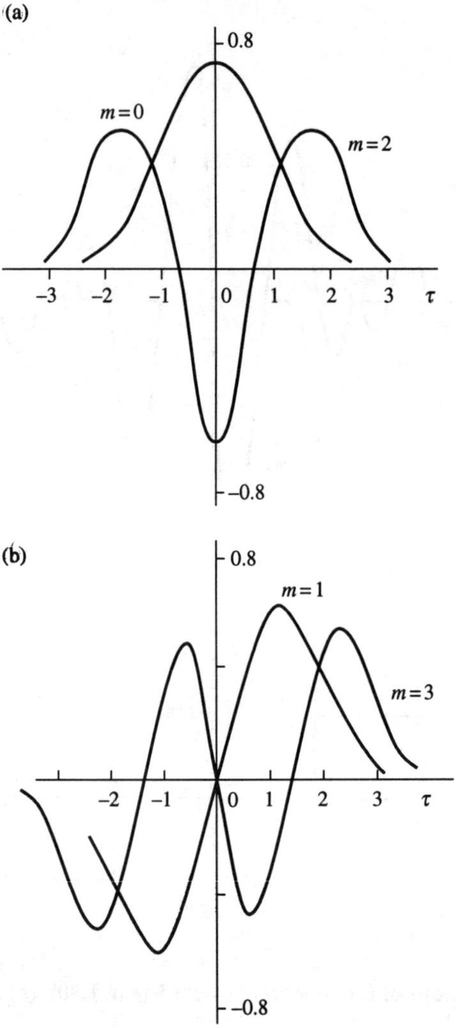

Fig. 1.9 (a) Even and (b) odd Hermitian pulses.

It is expedient to point out three properties of Hermitian pulses, important for optical applications:

1. The intervals between the zeros of Hermitian functions are unequal; thus these functions are suitable for description of oscillating waveforms with time-dependent frequencies.
2. The oscillations of the wings ($\tau \gg 1$) of Hermitian waveforms with high m ($m \gg 1$) are close to sinusoidal oscillations with decreasing amplitude (Ryzhik and Gradshtein 1980):

$$\Phi_{2m}(\tau)|_{\tau\gg 1} = C_{2m}(-1)^m 2^m (2m-1)!!\cos[\sqrt{(4m+1)}\tau], \quad (1.32)$$

$$\Phi_{2m+1}(\tau)|_{\tau\gg 1} = C_{2m+1}(-1)^m 2^{m+1/2}(2m-1)!!\sqrt{(2m+1)}\sin[\sqrt{(2m+3)}\tau]. \quad (1.33)$$

3. The spectra of Hermitian functions $(\Phi_m)_\omega$ are continuous and analogous to these functions, e.g.:

$$(\Phi_1)_\omega = \mathrm{i}\sqrt{2\pi}\,\Phi_1(u), \quad (1.34)$$

$$(\Phi_2)_\omega = -2\pi\,\Phi_2(u), \quad (1.35)$$

$$u = \omega T.$$

Proceeding in a similar fashion to that in Subsection 1.2.1, one can find the spectral bandwidths k_m of the pulses Φ_m; thus the values k_1 and k_2 for the pulses Φ_1 and Φ_2 are $k_1 = 2.35$, $k_2 = 2.8$.

As compared with Walsh signals and wavelets shown in Fig. 1.8, the Hermitian envelopes possess some of their advantages and are free from their shortcomings; these envelopes are continuous, like the wavelets eqn (1.28), but their spectra are also continuous like the spectra of Walsh signals.

The Hermitian functions can be used for expanding arbitrary waveforms throughout the time domain ($-\infty < \tau < \infty$). Thus it is easy to see from eqn (1.29) that one of the widely used models of wavelets, the so-called Mexican-hat wavelet (Fig. 3.2) introduced by Gabor (1946):

$$F = (1-\tau^2)\exp(-\tau^2/2), \quad (1.36)$$

can be expressed by means of Hermitian pulses Φ_0 and Φ_2:

$$F(\tau) = \frac{1}{2C_0}\,[\Phi_0(\tau) - \sqrt{2}\,\Phi_2(\tau)] \quad (1.37)$$

The Hermitian pulses will be shown in Chapter 3 to form one of the exactly solvable models of shape-dependent interactions of ultrashort non-sinusoidal waveforms with dispersive and lossy media.

1.3 Nonlinear geometrical optics of pulses with time-dependent frequencies

When considering, above, the spectral properties of solitary waveforms, we supposed the medium of propagation to be lossless and stationary, hence the spectra of waveforms remained invartant in this model. However, the propagation of EM fields through a nonstationary medium is known to provide a temporal variation of the spectral content of these fields. In the case of slowly varying parameters of transparent homogeneous media, when the time scales of such variations are much longer than the periods of the travelling sinusoidal waves, one can find the phase modulation of these waves φ by means of time-dependent perturbation $\Delta n(t)$ of refractive index $n = n_0 + \Delta n(t)$:

$$\varphi(t) = \frac{\omega_0 z}{c} \Delta n(t). \tag{1.38}$$

Here ω_0 is the carrier frequency and z is the length of wave path. This phase effect results in the spectral perturbation

$$\Delta\omega = -\frac{\partial \varphi(t)}{\partial t}. \tag{1.39}$$

In the quasi-stationary limit discussed, the perturbation $\Delta\omega$ is relatively small: $|\Delta\omega|\omega_0^{-1} \ll 1$.

Putting aside the general problem of power-dependent dynamics of solitary pulses, let us consider here the time-dependent modulation of frequencies of such pulses and controlled distribution of this modulation along the envelope.

An important way of tuning the spectral content of the pulse is associated with the formation of a time-dependent perturbation of the refractive index by means of nonlinear phenomena. Numerous materials used in nonlinear all-optical systems are characterized by third-order nonlinearity due to the Kerr effect, when the electric displacement $\boldsymbol{D}$ may be presented as (Stegeman *et al.* 1988)

$$\boldsymbol{D} = \varepsilon_0 n_0^2 \boldsymbol{E} + \boldsymbol{D}_n \tag{1.40}$$

Here ε_0 is the permittivity of free space and n_0 is the unperturbed refractive index. The power-dependent part of the electric displacement $\boldsymbol{D}_n$ may be written in the form

$$\boldsymbol{D}_n = \varepsilon_0 \chi [A|\boldsymbol{E}|^2 \boldsymbol{E} + B(\boldsymbol{E}^*)^2 \boldsymbol{E}] \tag{1.41}$$

where χ is a nonlinear constant and A and B are dimensionless coefficients characterizing the particular mechanisms of nonlinearity. In a lossless medium, A and B are real quantities and

$$A + B = 1, \tag{1.42}$$

The value BA^{-1} depends on the properties of the material and the mechanism of nonlinearity: thus $BA^{-1} = 3$ in liquid dielectrics with strong power-dependent polarizability, e.g. CCl_4; $BA^{-1} = 0.5$ in some glasses with electron nonlinearity, the influence of electric field on the orientation of molecules in the vitreous state being taken into account (Paillette 1987; Trillo *et al.* 1989); finally, $B = 0$ for electrostriction. Typical values of the nonlinear Kerr constant χ for the series of materials utilized in integrated and waveguide optics are small: 1 to $3 \times 10^{-22}\,\mathrm{m^2\,V^{-2}}$ for silica fibres, and 0.5 to $1 \times 10^{-24}\,\mathrm{m^2\,V^{-2}}$ for liquid CS_2.

If the pulse duration T is much longer than the relaxation time T_0 of nonlinear response of the medium, one can evaluate the self-induced spectral perturbations of a pulse produced by its envelope $E(t)$. Considering, for simplicity, the case $B = 0$, $A = 1$, and making use of the well-known relation between the refractive index n and the permittivity of the medium, $\varepsilon = n^2$, we can link the power-dependent perturbation of refractive index with the Kerr constant, eqn (1.41):

$$\Delta n(t) = \frac{\varepsilon_0 \chi |E(t)|^2}{2n_0}. \tag{1.43}$$

Substituting eqn (1.43) in eqns (1.38)–(1.39), we find the profiles of phase-shift $\varphi(t)$ and frequency modulation $\Delta\omega(t)$ along the pulse, the envelope $E(t)$ being given. Thus the spectra of some pulses in a nonlinear medium depend on the distance of propagation.

However, this consideration is valid only so long as the frequency variation has no influence on the electromagnetic field $E(t)$, perturbing the refractive index. Such a situation can arise in some limiting cases, e.g. when the distributions $\varphi(t)$ and $\Delta\omega(t)$ relate to a weak pulse propagating in the medium, perturbed by a strong pulse $E(t)$, eqn (1.43), or when the dispersion of the medium proves to be negligible.

To analyse the more general case, a self-consistent consideration of coupled nonlinear variations of spectral and amplitude envelopes of the pulse is required. The equation governing these coupled variations may be derived from the Maxwell equations:

$$\operatorname{rot} \boldsymbol{E} = -\frac{1}{c}\frac{\partial \boldsymbol{B}}{\partial t}, \tag{1.44}$$

$$\operatorname{rot} \boldsymbol{H} = \frac{1}{c}\frac{\partial \boldsymbol{D}}{\partial t}, \tag{1.45}$$

$$\operatorname{div} \boldsymbol{B} = 0, \tag{1.46}$$

$$\operatorname{div} \boldsymbol{D} = 0, \tag{1.47}$$

where $\boldsymbol{H}$ and $\boldsymbol{E}$ are the magnetic and electric fields and $\boldsymbol{B}$ and $\boldsymbol{D}$ are magnetic inductance and electric displacement, respectively.

In dealing with nonmagnetized media it is worth while to derive from the system of eqns (1.44)–(1.47) a single equation governing the behaviour of the electric field $\boldsymbol{E}$. Combining the two equations, (1.44) and (1.45), we obtain

$$\Delta \boldsymbol{E} - \text{grad div } \boldsymbol{E} - \frac{1}{c^2}\frac{\partial^2 \boldsymbol{D}}{\partial t^2} = 0. \tag{1.48}$$

Here Δ is the Laplace operator and c is the velocity of light in free space; the magnetic properties of the medium are those of a vacuum: $\boldsymbol{B} = \mu_0 \boldsymbol{H}$, where μ_0 is the magnetic permeability of vacuum. The electric displacement $\boldsymbol{D}$, defined by eqn (1.40), can be expressed via the dielectric permittivity of the medium:

$$\left.\begin{aligned} \boldsymbol{D} &= \varepsilon(\omega, t, |\boldsymbol{E}|^2)\boldsymbol{E} \\ \varepsilon(\omega, t, |\boldsymbol{E}|^2) &= \varepsilon(\omega, t) + \chi|\boldsymbol{E}|^2 \end{aligned}\right\} \tag{1.49}$$

The term $\varepsilon(\omega, t)$ in eqn (1.49) describes the variation of dielectric permittivity.

Let us consider in this section the pulses of transversely polarized time-harmonic waves travelling in a nonlinear lossless medium in the z-direction. The term grad div $\boldsymbol{E}$ in eqn (1.48) vanishes when there is only one electric component in the electromagnetic field discussed. Such a polarization structure occurs for example in the following wave problems:

(1) propagation of a plane wave in an unbounded medium;

(2) propagation of TE_{0p} modes in a slab waveguide;

(3) propagation of axisymmetrical modes in circular fibres.

Keeping in mind these structures, one can represent the solution of the low-intensity limit of eqn (1.49) ($\boldsymbol{D} = \varepsilon_0 n_0^2 \boldsymbol{E}$) in the form of a travelling plane wave:

$$E = E_0 F(t) \exp[\mathrm{i}(\beta z - \omega t)]. \tag{1.50}$$

Here $F(t)$ is the pulse envelope and β is the wavenumber:

$$\beta = \frac{\omega n_0}{c}. \tag{1.51}$$

To derive a self-consistent equation governing the coupled evolution of amplitude and spectra of the solitary pulse, let us rewrite the field of eqn (1.51) in a generalized form:

$$E_{0p} = \Psi_{0p}(x) F(z, t) \int \mathrm{d}\omega\, E_0(\omega) \exp\{\mathrm{i}[\beta(\omega) z - \omega t]\}. \tag{1.52}$$

Here the function $\Psi_{0p}(x)$, which is an eigenfunction of the $0p$-th mode, determines the field distribution in a plane orthogonal to the z-axis; x is the coordinate in this plane. The spatial and temporal evolution of the pulse envelope is described by an unknown complex amplitude $F(z, t)$. This function can be found by substitution of eqns (1.41) and (1.52) in the Maxwell equation (1.48) and expanding the frequency-dependent dielectric permittivity $\varepsilon(\omega)$ in a Taylor series up to second order in the carrier frequency ω:

$$\varepsilon(\omega)=\varepsilon(\omega_0)+(\omega_0-\omega)\frac{\partial\varepsilon}{\partial\omega}+\frac{1}{2}(\omega_0-\omega)^2\frac{\partial^2\varepsilon}{\partial\omega^2} \tag{1.53}$$

We shall follow the method of slowly varying amplitudes widely used in nonlinear electromagnetic field theories. Recognizing that each power of $(\omega_0-\omega)$ in the formula (1.53) is equivalent to $\mathrm{i}\partial/\partial t$ in the sum

$$\varepsilon(\omega,t)=\varepsilon_0(\omega_0)+\mathrm{i}\frac{\partial\varepsilon}{\partial\omega}\frac{\partial}{\partial t}-\frac{1}{2}\frac{\partial^2\varepsilon}{\partial\omega^2}\frac{\partial^2}{\partial t^2}, \tag{1.54}$$

one can rewrite the last term in eqn (1.48) in the form

$$-\frac{1}{c^2}\frac{\partial^2 D}{\partial t^2}=\left[\frac{\omega^2\varepsilon}{c^2}E+\frac{2\mathrm{i}\beta}{v_g}\frac{\partial E}{\partial t}-\frac{1}{v_g^2}(1-\beta v_\omega)\frac{\partial^2 E}{\partial t^2}\right] \tag{1.55}$$

Here v_ω is the derivative of the group velocity v_g:

$$v_g=\frac{\partial\omega}{\partial\beta},\quad v_\omega=\frac{\partial v_g}{\partial\omega}. \tag{1.56}$$

Substituting eqns (1.52) and (1.55) in eqn (1.48), we obtain a self-consistent equation governing the pulse dynamics in a nonlinear dispersive lossless medium. It is expedient to present this equation in a normalized form:

$$\mathrm{i}\frac{\partial F}{\partial\eta}+\frac{1}{2}\frac{\partial^2 F}{\partial q^2}-\chi_{0p}|F|^2F=0. \tag{1.57}$$

Here the dimensionless function F is the complex envelope of the pulse of eqn (1.52) and the variables η and q are

$$\eta=\frac{z}{L_\omega},\quad \mathrm{q}=\frac{t-zv_g^{-1}}{T}, \tag{1.58}$$

where L_ω is the dispersive length, related to the characteristic duration of the pulse T by

$$L_\omega=\frac{(v_g T)^2}{v_\omega}, \tag{1.59}$$

v_ω being defined by eqn (1.56). The nonlinear parameter is χ_{0p} determined for each $0p$-th mode:

$$\chi_{0p}=\frac{\omega^2}{2c^2}\frac{L_\omega}{\beta}\chi|E_0|^2G_{0p}. \tag{1.60}$$

The values of wavenumber β, dispersive length L_ω and amplitude E_0 in eqn (1.60) characterize the $0p$-th mode. The coefficient G_{0p} in eqn (1.60), which appears as a result of averaging of the heterogeneous field of the $0p$-th mode over the cross section of the wave beam, may be written as a fraction:

$$G_{0p} = \frac{\int |\Psi_{0p}(x)|^4 \, \mathrm{d}s}{\int |\Psi_{0p}(x)|^2 \, \mathrm{d}s}. \tag{1.61}$$

In the limiting case of propagation of a homogeneous plane wave in an unbounded medium, $G = 1$.

Equation (1.57) is the simplest form of so-called nonlinear Schrödinger equation, providing a self-consistent description of coupled amplitude-frequency modulation of a pulse of sinusoidal waves in a nonlinear dispersive medium. This equation is widely used in a multitude of nonlinear problems for investigation of self-modulated and self-localized states in systems of different physical nature. However, explicit analytical solutions of eqn (1.57), describing the power-dependent evolution of ultrashort pulses in dispersive media, are unknown. Some trends of such a process may be revealed approximately by means of a variational approach (Shvartsburg 1982), based on a hypothesis of self-similar evolution of Gaussian envelopes. The overwhelming majority of results describing the nonstationary propagation of solitary pulses in the framework of eqn (1.57) have so far been obtained by computer similation.

The stationary solution of this equation in the case $\chi > 0$ is the so-called fundamental soliton:

$$F = \frac{\exp(\mathrm{i}\eta\Delta^{-2})}{\mathrm{ch}(q/\Delta)}, \quad \Delta = \chi^{-1/2}. \tag{1.62}$$

The shape and peak intensity of these stationary pulses are such that effects of nonlinearity always exactly cancel the dispersive broadening; the width and shape of the pulse, eqn (1.62), therefore remain constant during propagation, as well as its Fourier spectrum

$$F_\omega = \frac{\pi T}{\mathrm{ch}(0.5\pi\omega t)}. \tag{1.63}$$

Thus, from the standpoint of pulse shaping, the soliton envelope may be considered as an exceptional one due to distortion-free stationary propagation in a dispersive medium. There is a vast literature on the soliton concept and its numerous applications; see for example Zakharov *et al.* (1980), Jeunhomme (1993) and the references therein.

On the other hand, attention is focused below on nonstationary propagation of pulses, providing controlled shaping and spectral tuning. To develop the analytical theory of such coupled amplitude–frequency modulation it is worthwhile to present the complex amplitude of the pulse F in the form

$$F(\tau, q) = f(\tau, q) \exp[is(\tau, q)]. \tag{1.64}$$

Substitution of eqn (1.64) in eqn (1.57) and separation of the real and imaginary parts of eqn (1.57) permits one to reduce this equation to a pair of coupled equations governing the dimensionless intensity W and frequency modulation v:

$$W = f^2, \quad v = \frac{1}{\sqrt{\chi}} \frac{\partial s}{\partial q} \operatorname{sign} v_\omega. \tag{1.65}$$

Here the function sign x is

$$\operatorname{sign} x = \begin{cases} +1, & x > 0, \\ -1, & x < 0. \end{cases} \tag{1.66}$$

One of these equations relates to the real part of (1.57) and may be written as

$$\frac{\partial v}{\partial \tau} + v \frac{\partial v}{\partial q} + \left[\frac{\partial W}{\partial q} + \frac{1}{\chi_{0p}} \frac{\partial}{\partial q} \left(\frac{1}{\sqrt{W}} \frac{\partial^2 \sqrt{W}}{\partial q^2}\right)\right] \operatorname{sign}(\chi v_\omega) = 0. \tag{1.67}$$

The dimensionless distance τ is related to the spatial variable η, eqn (1.58), by

$$\tau = \eta \sqrt{|\chi_{0p}|}, \tag{1.68}$$

the variable q being given, as before, by eqn (1.58).

The nonlinear Schrödinger equation (1.57) contains one parameter χ_{0p}. The values of this parameter for optical power-carrying waveguide systems are usually in the region of 1 to 3×10^2. The presence of a large factor

$$|\chi_{0p}| \gg 1 \tag{1.69}$$

In the denominator of the last term in eqn (1.67) permits one to neglect this term, at least outside the range of small intensities W. Omitting this term in eqn (1.67) and using another equation, obtained without any new assumptions about the imaginary part of eqn (1.57), one may reduce eqn (1.57) to the following set of equations:

$$\frac{\partial v}{\partial \tau} + v \frac{\partial v}{\partial q} + \frac{\partial W}{\partial q} \operatorname{sign}(\chi v_\omega) = 0, \tag{1.70}$$

$$\frac{\partial W}{\partial \tau} + \frac{\partial (Wv)}{\partial q} = 0. \tag{1.71}$$

These are the fundamental equations of nonlinear, nonstationary geometrical optics. They describe the coupled distributions of intensity W and frequency modulation v along the pulse. The initial conditions for this system of equations are given by the intensity and frequency envelopes of the input pulse:

$$W|_{\tau=0} = W_0(q), \quad v|_{\tau=0} = v_0(q). \tag{1.72}$$

The system of nonlinear equations (1.70)–(1.71) has wide classes of exact analytical solutions. To analyse these solutions it is expedient to interchange the independent variables q, τ and the functions $W(\tau, q)$ and $v(\tau, q)$, considering the new functions

$$q = q(W, v), \quad \tau = \tau(W, v). \tag{1.73}$$

This interchange reduces the nonlinear system (1.70)–(1.71) to a linear one. Representing the derivatives in eqns (1.70)–(1.71) by means of Jacobians, e.g.

$$\frac{\partial v}{\partial \tau} \to \frac{\mathrm{D}(v, \eta)}{\mathrm{D}(\tau, \eta)},$$

and multiplying, the results of such representation on Jacobian $\partial(\tau, \eta)/\partial(W, v)$, we obtain the following new form of fundamental system of equations of nonlinear, nonstationary geometrical optics:

$$\frac{\partial q}{\partial W} - v\frac{\partial \tau}{\partial W} + \frac{\partial \tau}{\partial v}\,\mathrm{sign}(\chi v_\omega) = 0, \tag{1.74}$$

$$\frac{\partial q}{\partial v} + W\frac{\partial \tau}{\partial W} - v\frac{\partial \tau}{\partial v} = 0. \tag{1.75}$$

Similarly to trivial geometrical optics, the wavelength of the radiation tends to zero in this approach. It is important to point out that, unlike linearization with respect to any small parameter, the transformed system (1.74)–(1.75) is exactly linear.

The solutions of the system (1.74)–(1.75) depend essentially on the sign of the product χv_ω, eqn (1.70), characterizing the interplay of dispersive and nonlinear properties in a Kerr-like medium. This interplay can be responsible for opposite tendencies in pulse dynamics, manifested in the processes of self-compression ($\chi v_\omega < 0$) and self-steepening ($\chi v_\omega > 0$) of the pulse. Hence the cases $\chi v_\omega < 0$ and $\chi v_\omega > 0$, embracing four combinations of positive and negative nonlinearities with normal and abnormal dispersion, are analysed separately below, and the possibilities of controlled tuning of time-dependent frequency modulation are analysed for both of these cases.

1.3.1 Controlled nonstationary spectral broadening close to the peak of the envelope

Let us consider here the amplitude–frequency modulation of a pulse in the range $\chi v_\omega < 0$, relevant in particular to the propagation of powerful infrared radiation with wavelengths $\lambda > 1.3\,\mu$m in a silica fibre (abnormal dispersion, $v_\omega < 0$; positive nonlinearity, $\chi > 0$). This combination of parameters, providing the self-confinement of solitary pulses, is known to be important for long-distance optical communication systems (Mollenauer *et al.* 1985).

To obtain exact analytical solutions of the system (1.74)–(1.75) it is useful to reduce thls system to one differential equation of second order, expressing the functions $\tau(W, v)$ and $q(W, v)$ by a new function

$$\tau = \frac{\partial \Psi}{\partial W}\,\mathrm{sign}\,(\chi v_\omega), \tag{1.76}$$

$$q = v\tau - \frac{\partial \Psi}{\partial v}. \tag{1.77}$$

Substitution of eqns (1.76)–(1.77) in the fundamental system (1.74)–(1.75) reduces eqn (1.74) to an identity, whereas eqn (1.75) yields the equation governing the function Ψ in (W, v)-space:

$$W\frac{\partial^2 \Psi}{\partial W^2} + \frac{\partial \Psi}{\partial W} - \frac{\partial^2 \Psi}{\partial v^2}\,\mathrm{sign}\,(\chi v_\omega) = 0. \tag{1.78}$$

Considering for simplicity the case of an initially unmodulated wave $v_0(q) = 0$, and combining eqns (1.72) and (1.76)–(1.77), we can write the boundary conditions to eqn (1.79) at a plane $v = 0$ *in* (W, v)-space as

$$\left.\frac{\partial \Psi}{\partial W}\right|_{v=0} = 0, \tag{1.79}$$

$$\left.\frac{\partial \Psi}{\partial v}\right|_{v=0} = -q_0(W), \tag{1.80}$$

where $q_0(W)$ is the function inverse to $W_0(q)$, eqn (1.72). Basically, the function determined by the second-order differential equation (1.78) with the boundary conditions (1.79)–(1.80) contains all the information about the power-dependent dynamics of the pulse. To provide flexibility of initial pulse envelopes $W_0(q)$, eqn (1.72), described by the boundary conditions (1.79)–(1.80), it is useful to introduce the ellipsoidal coordinate system (ε, ξ) in (W, v)-space:

$$W = (1 + \varepsilon^2)(1 - \xi^2), \quad v = 2\varepsilon\xi. \tag{1.81}$$

This coordinate system is shown in Fig. 1.10; the plane $v = 0$ relates to the disk $\varepsilon = 0$. Finding from eqn (1.81) the derivatives

$$\frac{\partial \Psi}{\partial W} = -\frac{1}{2(\varepsilon^2 + \xi^2)}\left(\varepsilon\frac{\partial \Psi}{\partial \varepsilon} - \xi\frac{\partial \Psi}{\partial \xi}\right), \tag{1.82}$$

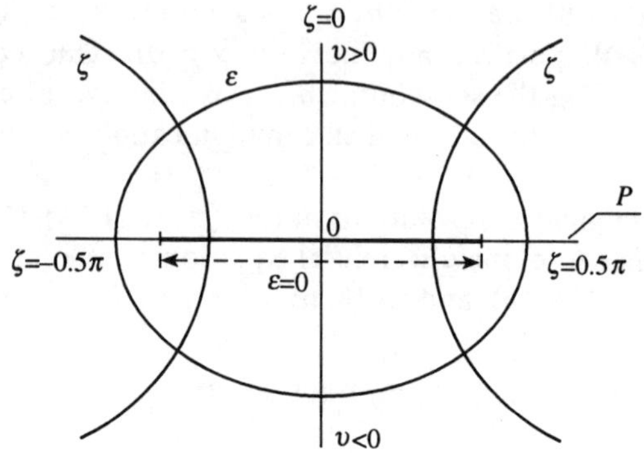

Fig. 1.10 Axisymmetrical ellipsoidal ($\varepsilon = \text{const.}$) and hyperboloidal ($\xi = \text{const.}$) coordinate surfaces, orthogonal in (W, v)-space. The boundary conditions (1.79)–(1.80) and (1.85)–(1.86) are given on the planes $v = 0$ and $\varepsilon = 0$ respectively.

$$\frac{\partial \Psi}{\partial v} = \frac{1}{2(\varepsilon^2 + \xi^2)} \left(\xi(1 + \varepsilon^2) \frac{\partial \Psi}{\partial \varepsilon} + \varepsilon(1 - \xi^2) \frac{\partial \Psi}{\partial \xi} \right), \tag{1.83}$$

we may rewrite eqns (1.79)–(1.81) in the form (Shvartsburg 1974)

$$\frac{\partial}{\partial \varepsilon}\left[(1 + \varepsilon^2) \frac{\partial \Psi}{\partial \varepsilon}\right] + \frac{\partial}{\partial \xi}\left[(1 - \xi^2) \frac{\partial \Psi}{\partial \xi}\right] = 0. \tag{1.84}$$

Here

$$\left.\frac{\partial \Psi}{\partial \varepsilon}\right|_{\varepsilon=0} = -2\xi q_0(\xi), \quad \left.\frac{\partial \Psi}{\partial \xi}\right|_{\varepsilon=0} = 0, \tag{1.85}$$

$$\xi|_{\varepsilon=0} = \sqrt{1 - W}. \tag{1.86}$$

Equation (1.84) is the well-known three-dimensional axisymmetrical Laplace equation; using its eigenfunctions P_n and Q_n, one may write the general solution of eqn (1.85) in the form

$$\Psi = \sum_{n=0}^{\infty} \{Q_n(\xi)[A_n Q_n(\mathrm{i}\varepsilon) + B_n P_n(\mathrm{i}\varepsilon)] + P_n(\xi)[C_n Q_n(\mathrm{i}\varepsilon) + D_n P_n(\mathrm{i}\varepsilon)]\}. \tag{1.87}$$

The eigenfunctions $P_n(\xi)$, $Q_n(\xi)$, $P_n(\mathrm{i}\varepsilon)$ and $Q_n(\mathrm{i}\varepsilon)$ are given in Appendix 1. The unknown coefficients A_n, B_n, C_n, D_n are determined by substitution of eqn (1.87) in the boundary conditions (1.85). The evolution of the initial pulse is now described by substitution of function Ψ, eqn (1.87) in the correlations (1.76)–(1.77) and calculation of the relevant derivatives by means of eqns (1.82)–(1.83).

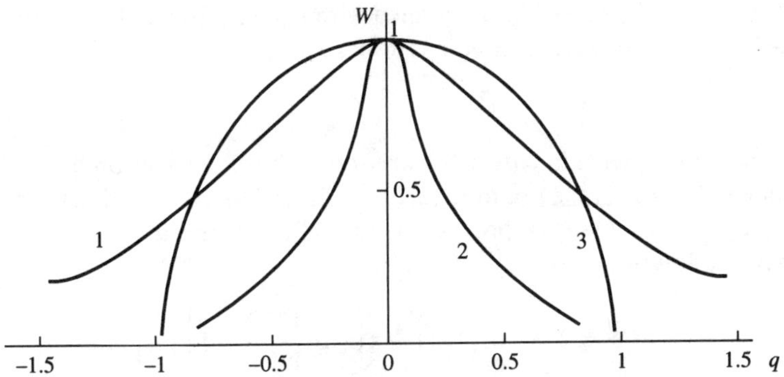

Fig. 1.11 Initial intensity profiles $W_0(q)$ of pulses with coincident peak value $W_{max} = 1$ and different wings: curve 1, the (hyperbolic-secant)^2 envelope (eqn 1.88); curves 2 and 3, belonging to the one-parameter family of envelopes (eqn 1.92), relate to $a = -1.1$ and $a = -10$ respectively.

Thus all the information, regarding amplitude–frequency self-modulation of the initial pulse in a Kerr-like dispersive medium is contained in the function (1.88). This approach provides a simple standardized scheme for exact analytical solution of equations of nonlinear geometrical optics for wide classes of initial pulses $W_0(q)$, when the inverse dependence $q_0(W)$, characterizing the initial intensity envelope, is single-valued. Some possibilities of this method in analysis of controlled sell-modulation of pulses are illustrated below.

(A) Flexibility of initial conditions An important feature of nonlinear geometric optics (NGO) is the possibility of varying the initial envelope, adding complementary harmonics to the sums (1.87) or changing the contribution of some harmonics due to variations of the relevant coefficients in eqn (1.87). To visualize this flexibility, let us begin from the simplest and widely used (hyperbolic-secant)^2 envelope (Fig. 1.11). Keeping in mind the condition (1.86), we obtain

$$W_0 = \mathrm{ch}^{-2} q, \quad q_0 = \frac{1}{2} \ln \frac{1+\xi}{1-\xi}. \tag{1.88}$$

Using the formulae for eigenfunctions $Q_n(\xi)$ and $P_n(\xi)$ of Laplace equation (1.87) (see Appendix 1) and substituting q_0 from eqn (1.88) in the boundary conditions (1.85), one can rewrite the first of these conditions as

$$\left.\frac{\partial \Psi}{\partial \varepsilon}\right|_{\varepsilon=0} = -2Q_1(\xi) - 2P_0(\xi). \tag{1.89}$$

The second condition (1.88), combined with eqn (1.89), yields the following values of the coefficients in eqn (1.87):

$$B_1 = -2, \quad C_0 = 2. \tag{1.90}$$

The other coefficients in eqn (1.87) are zero. Thus the function $\Psi(\varepsilon, \xi)$ and the dependences $\tau(\varepsilon, \xi)$ and $q(\varepsilon, \xi)$ eqns (1.76)–(1.77), describing the reshaping of the intensity profile, eqn (1.88), and the appearance of frequency modulation, are

$$\left.\begin{aligned} \Psi &= 2\left[\operatorname{arctg}(1/\varepsilon) - \varepsilon\left(\frac{\xi}{2}\ln\frac{1+\xi}{1-\xi} - 1\right)\right], \\ \tau &= \frac{\varepsilon}{(1+\varepsilon^2)(1-\xi^2)}, \\ q &= \frac{1}{2}\ln\frac{1+\xi}{1-\xi} - \frac{2\xi\varepsilon^2}{(1+\varepsilon^2)(1-\xi^2)}. \end{aligned}\right\} \tag{1.91}$$

However, the model of the initial profile, eqn (1.88), illustrating the sequence of calculations for determination of the function Ψ, has two shortcomings impeding its use for interpretation of real optical experiments:

(1) the profile (1.88) is defined in an infinite time domain $(-\infty < q < \infty)$;

(2) there is no free parameter in the envelope $W_0 = \mathrm{ch}^{-2} q$.

To eliminate both of these shortcomings, let us consider the more complex envelope

$$q_0 = \frac{1-\xi^2}{2a}\ln\frac{1+\xi}{1-\xi} + \xi. \tag{1.92}$$

This bell-like profile, containing one free parameter a,

$$a < -1, \tag{1.93}$$

is determined in a limited time domain $(-1 \leqslant q \leqslant 1)$ (Fig. 1.11). The wings of this pulse tend to zero smoothly:

$$W_0|_{q=\pm 1} = 0, \quad \left.\frac{\partial W_0}{\partial q}\right|_{q=\pm 1} = 0. \tag{1.94}$$

Variations in parameter a can provide changes in width of the envelope, eqn (1.92), with its peak amplitude $(W_m = 1)$ and the behaviour at the edges (1.94) invariant. The limit $a \to -\infty$ relates to a parabolic shape (Fig. 1.11). The coefficients in the presentation of function Ψ, eqn (1.87), related to eqn (1.92) and calculated by analogy with eqn (1.89), are

$$\left.\begin{aligned} A_0 = 2/3(1+1/a), \quad B_1 = -4/(5a), \quad C_2 = 2/3(1-1/a), \\ D_2 = \pi/3(1/a-1), \quad B_3 = 8/(15a). \end{aligned}\right\} \tag{1.95}$$

If the flexibility of the initial envelope, eqn (1.92), provided by one free parameter a proves to be insufficient for representation of any pulse, the model (1.92) may be improved by adding other eigenfunctions of eqn (1.87). Thus, one can use an initial envelope with two parameters a and b:

$$q_0 = (a+b)^{-1}\left[\frac{1-\xi^2}{2}\ln\frac{1+\xi}{1-\xi} + a\xi + b\xi^3\right]. \tag{1.96}$$

The coefficients a and b obey the conditions

$$1 + a + 3b < 0, \quad a + b \neq 0. \tag{1.97}$$

The pulse (1.96) is localized in the time domain $(-1 \leqslant q < 1)$, and the conditions at the edges (1.94) are fulfilled.

These procedures can be continued, improving the approximation of the initial profile.

(B) Non-stationary modulation of intensity and spectral envelope Considering the 'inverse' dependences $\tau = \tau(W, v)$ and $q = q(W, v)$ obtained by substitution of function Ψ, eqn (1.87) in eqns (1.79)–(1.80), we cannot usually restore the explicit formulae for $W = W(\tau, q)$ and $v = (\tau, q)$. However, the evolution of the pulse may be analysed by means of these 'inverse' functions through the following scheme of calculations.

To determine the intensity and spectral envelope of the pulse at some point $\tau = \tau_0(\varepsilon, \xi)$ inside the medium, one must find the values ε as the function of ξ, the distance τ_0 being given. Since the travelling coordinate $q(\varepsilon, \xi)$ for symmetrical pulses (Fig. 1.11) was shown to be an antisymmetric function of the variable $\xi[q(\xi) = -q(-\xi)]$, it is enough to analyse the dependence $\varepsilon = \varepsilon(\tau_0, \xi)$ only in the range $0 \leqslant \xi \leqslant 1$. Here, the peak intensity W_m for each distance τ_0 relates to the value $\xi = 0$:

$$W_m = 1 + \varepsilon_m^2, \quad \varepsilon_m = \varepsilon(\tau_0)\,|_{\xi=0} \tag{1.98}$$

Using the values ξ and $\varepsilon = \varepsilon(\xi, \tau_0)$, one can find the reshaping of the intensity profile and frequency modulation, the so-called 'chirp' v, eqn (1.65), relevant to each value $q = q(\varepsilon, \xi)$ for the given point τ_0. Substituting the function Ψ, eqn (1.87), with the coefficients (1.95) in eqns (1.82)–(1.83) and making use of definitions (1.76)–(1.77), we finally obtain the solution governing the evolution of the initial pulse (1.92):

$$\tau = \frac{1}{2}\left(1 - \frac{1}{a}\right)\left(\operatorname{arctg}\frac{1}{\varepsilon} - \frac{\pi}{2} - \frac{\varepsilon}{1+\varepsilon^2}\right)$$

$$+ \frac{2\varepsilon}{a}\left[\frac{\xi}{2}\ln\frac{1+\xi}{1-\xi} - \frac{1}{1+\varepsilon^2} + \frac{\varepsilon^2(\xi^2 - 23^{-1})}{(1+\varepsilon^2)(1-\xi^2)}\right], \tag{1.99}$$

$$q = \frac{\xi}{1+\xi^2} + \frac{1}{a}\left[\frac{1+\varepsilon^2+\varepsilon^2\xi^2-\xi^2}{2}\ln\frac{1+\xi}{1-\xi} + \frac{\xi\varepsilon^4(1+3\xi^2)}{3(1+\varepsilon^2)(1-\xi^2)}\right]. \tag{1.100}$$

The envelopes $W = W(q, \tau)$ and $v = v(q, \tau)$ formed as a result of the nonlinear evolution of the original pulse (1.95) are shown in Fig. 1.12.

The time-dependent modulation $\Delta\omega(t)$ of carrier frequency ω_0 is determined from eqns (1.65) and (1.39):

$$\Delta\omega(t) = \frac{v}{T}\sqrt{|\chi_{0p}|}\ \operatorname{sign}(\chi v_\omega). \tag{1.101}$$

Keeping in mind the usual values $|v| \approx 1$, $|\chi_{0p}| \gg 1$, one can see that the nonlinear frequency 'chirp', eqn (1.102), may exceed by $(|\chi_{0p}|)^{1/2}$ times the spectral bandwidth of the bell-like pulse, dictated by its shape and time scale, eqn (1.1). This time-dependent frequency $\Delta\omega(t)$ is distributed along the pulse proportionally to the function $v(q)$ (Fig. 1.12b). It is important to point out that the nonstationary frequency envelope $\Delta\omega(t)$ formed in the course of coupled amplitude–frequency evolution is governed by the original profile of intensity.

The nonlinear evolution of the initial pulse profile (Fig. 1.11, curve 1) causes narrowing of the central part of the pulse, accompanied by an increase in its maximum (Fig. 1.12a). This process develops simultaneously with the frequency modulation. Such a nonstationary frequency 'chirp' is localized close to the pulse peak, where the slope of the envelope is increasing.

The spectral envelope, shown in Fig. 1.12b, relates to the case $v_\omega < 0$, $\chi > 0$. A change in sign of these parameters ($v_\omega > 0$, $\chi < 0$), the initial pulse profile remaining the same, leads to the formation of the same intensity envelope (Fig. 1.12a), but the spectral envelope changes its sign (Fig. 1.12b, dashed line).

(C) Limitations of nonlinear geometrical optics It was emphasised above that by omitting the term containing the second-order derivatives in eqn (1.67), we restrict our analysis to the case $|\chi_{0p}| \gg 1$. Moreover, one-to-one correspondence between the values τ, q and W, v is required for the use of Jacobians in the derivation of eqns (1.74)–(1.75). The conditions (1.93) and (1.97) determining the initial pulses are related to this requirement. However, this one-to-one correspondence, fulfilled at the beginning of nonlinear

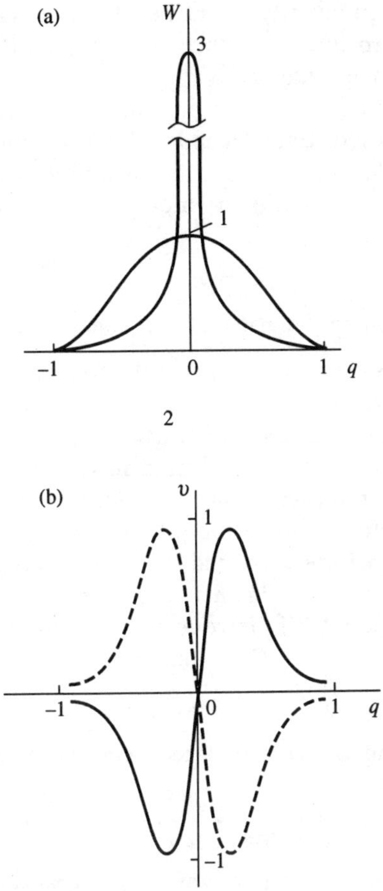

Fig. 1.12 Coupled nonlinear modulation of intensity and spectral envelopes at the central part of the pulse (eqn 1.92), $(a = -5)$, formed at the point $\tau = 0.5$ $(\chi v_\omega < 0)$: (a) the initial envelope of intensity W (eqn 1.92) and the self-compression of the pulse; (b) the solid and dotted lines show the distribution of the frequency 'chirp' along the pulse in the cases $\chi > 0$, $v_\omega < 0$ and $\chi < 0$, $v_\omega > 0$ respectively.

self-modulation, may be violated in the course of pulse propagation, owing to spike-like envelope formation. This process leads to the appearance of a region with an appreciable field gradient. In this region the derivative $\partial W/\partial q$ formally tends to infinity. Therefore the conditions of rigid intensity rise near the symmetric profile maximum may be written in the form

$$q = 0, \quad \left.\frac{\partial q}{\partial W}\right|_{\tau_{cr}} = 0. \tag{1.102}$$

The intensity at this singularity $\tau = \tau_{cr}$ has the finite value W_m, eqn (1.98); its derivative tends to infinity. When such singularities occur, the NGO equations are no longer valid.

Thus nonlinear geometric optics provides a unified and very common method for restoration of time-dependent intensity and frequency modulation of a short pulse localized close to its maximum. A simple generalization of this method for the nonstationary modulation of a pulse at its periphery is described below.

1.3.2 Time-dependent modulation of frequency at the pulse wings

It was illustrated in Subsection 1.3.1 that the frequency modulation accompanying the self-narrowing of the pulse intensity envelope ($\chi v_\omega < 0$) develops noticeably in the areas of increasing intensity gradient (Fig. 1.12b). The same tendency can be demonstrated in the framework of nonlinear geometrical optics for the evolution of pulses in the opposite case, $\chi v_\omega > 0$, characterizing for example the propagation of optical pulses in the visible region in silica fibres. To find the exact analytical solutions of the NGO equations (1.70)–(1.71) in this case, one can modify the approach advanced in Subsection 1.3.1, introducing instead of eqn (1.81) the new coordinate system (ε, ξ) in (W, v)-space:

$$W = (1 - \varepsilon^2)(1 - \xi^2), \quad v = -2\varepsilon\xi. \tag{1.103}$$

The Laplace equation (1.84) is replaced as a result of eqn (1.103) by

$$\frac{\partial}{\partial \varepsilon}\left[(1 - \varepsilon^2)\frac{\partial \Psi}{\partial \varepsilon}\right] - \frac{\partial}{\partial \xi}\left[(1 - \xi^2)\frac{\partial \Psi}{\partial \xi}\right] = 0. \tag{1.104}$$

The boundary conditions (1.79)–(1.80) are again given on the plane $\varepsilon = 0$. The solution of eqn (1.104) may be written by means of its eigenfunctions $P_n(\xi)$, $Q_n(\xi)$, $P_n(\varepsilon)$, $Q_n(\varepsilon)$ (see Appendix 1) in a form analogous to eqn (1.87). The eigenfunctions $P_n(\xi)$ and $Q_n(\xi)$ were utilized in eqn (1.87); the eigenfunctions $P_n(\varepsilon)$ and $Q_n(\varepsilon)$ are obtained from $P_n(\xi)$ and $Q_n(\xi)$ by the replacement $\xi \to \varepsilon$.

Proceeding in a similar fashion, we can derive the relevant modifications of all the formulae of Subsection 1.3.1. Omitting the interim calculations, let us consider the evolution of the bell-like pulse with finite duration, eqn (1.92), described in the case $\chi v_\omega > 0$ by the inverse dependences

$$\tau = +\frac{1}{2}\left(1 - \frac{1}{a}\right)\left[\frac{1}{2}\ln\frac{1+\varepsilon}{1-\varepsilon} + \frac{\varepsilon}{1-\varepsilon^2}\right]$$
$$-\frac{2\varepsilon}{a}\left[\frac{\xi}{2}\ln\frac{1+\xi}{1-\xi} - \frac{1}{1-\varepsilon^2} - \frac{\varepsilon^2(\xi^2 - 23^{-1})}{(1-\varepsilon^2)(1-\xi^2)}\right], \tag{1.105}$$

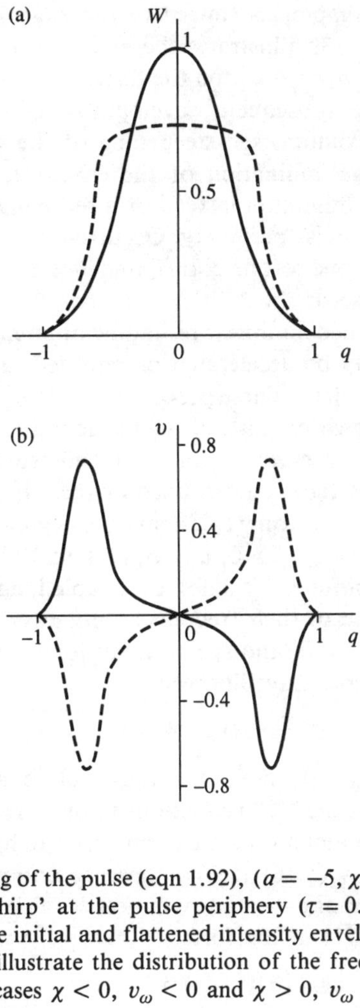

Fig. 1.13 Self-flattening of the pulse (eqn 1.92), ($a = -5, \chi v_\omega > 0$) and formation of a time-dependent 'chirp' at the pulse periphery ($\tau = 0.45$), (a) the solid and broken lines relate to the initial and flattened intensity envelope respectively (b) the solid and broken lines illustrate the distribution of the frequency 'chirp' close to the pulse edges in the cases $\chi < 0,\ v_\omega < 0$ and $\chi > 0,\ v_\omega > 0$ respectively.

$$q = \frac{\xi}{1-\xi^2} + \frac{1}{a}\left[\frac{1-\varepsilon^2-\xi^2-\varepsilon^2\xi^2}{2}\ln\frac{1+\xi}{1-\xi} + \frac{\xi\varepsilon^4(1+3\xi^2)}{3(1-\varepsilon^2)(1-\xi^2)}\right]. \quad (1.106)$$

Restoration of the explicit dependences $W = W(\tau, q)$ and $v = v(\tau, q)$, illustrated in Fig. 1.13, shows that the nonlinear spreading of the original pulse, eqn (1.92), in the course of propagation brings about flattening of its peak and an increase in pulse gradients at the wings of the envelope. This peripheral effect is known to give rise to the growth of a time-dependent frequency 'chirp' at the edges of the pulse. As compared with pulse, self-compression (Fig. 1.12), the maximum of the frequency chirp due to this pulse reshaping is shifted to the edges of the envelope. The dashed line in Fig. 1.13b illustrates the effect of a change in sign of the parameters, $\chi \to -\chi$, $v_\omega \to -v_\omega$, on the distribution of the frequency chirp along the pulse. The subsequent development of these tendencies (self-flattening of the maximum, self-steepening of the wings) will provide at some point either the lamination of the envelope, accompanied by the appearance of a new minimum instead of a pulse maximum, or formation of a 'spillover' at the pulse periphery. Occurrence of either of these phenomena will violate the one-to-one correspondence of the values of W, v and τ, q, impeding the use of the NGO equations.

The growth of coupled nonlinear reshaping of intensity and spectral envelope of the pulse may be accelerated or retarded, depending on the type of initial frequency 'chirp'. These processes are described, as before, by the Laplace equation; however, instead of on the plane $\varepsilon = 0$, eqn (1.85), the boundary conditions are given on some ellipsoidal surface $\varepsilon = \varepsilon_0$ in (W, v)-space. This surface is determined by the largest value of the modulation parameter $v = v_m$, eqn (1.65), at the beginning of the evolution, $\tau = 0$. Thus, in the case $\chi v_\omega > 0$, $\varepsilon_0 = v_m(1 + v_m^2)^{-1/2}$ (Shvartsburg 1993).

All the above-mentioned regimes of coupled nonlinear reshaping of pulses and modulation of their frequencies possess one common property: the profiles of intensity W and spectral envelope v cannot be represented in the form of a simple travelling wave,

$$W = W(v). \tag{1.107}$$

However, in the case $\chi v_\omega > 0$, one may obtain a new class of exact analytical solutions of the NGO equations in the form of solitary travelling waves, eqn (1.107), analogous to Riemann waves in hydrodynamics. Unlike the approach in eqns (1.76)–(1.77), these solutions provide an explicit description of the temporal–spatial evolution of initial pulses. Let us rewrite the system (1.70)–(1.71) in the form

$$\left.\begin{aligned} &\frac{\partial v}{\partial W}\left(\frac{\partial W}{\partial \tau} + v\frac{\partial W}{\partial q}\right) + \frac{\partial W}{\partial q} = 0, \\ &\frac{\partial W}{\partial \tau} + v\frac{\partial W}{\partial q} + v\frac{\partial v}{\partial W}\frac{\partial W}{\partial q} = 0. \end{aligned}\right\} \tag{1.108}$$

These equations are compatible with eqn (1.107) if the following condition is fulfilled:

$$\left.\frac{\partial W}{\partial \tau}\right|_W - \theta(W)\left.\frac{\partial W}{\partial q}\right|_W = 0. \tag{1.109}$$

Here $\theta(W)$ is an arbitrary function. Equation (1.109) shows that the values of intensity W are constant along the characteristics in the (τ, q)-plane, therefore

$$\frac{\partial q}{\partial \tau} = \theta, \quad q = \tau\theta + P(\theta). \tag{1.110}$$

The function $P(\theta)$ is determined by an initial profile

$$q_0 = q|_{\tau=0} = P(\theta). \tag{1.111}$$

Making use of eqn (1.110), one can derive from the set (1.108)

$$(v+\theta)\frac{\partial v}{\partial W} + 1 = 0, \quad v + \theta + W\frac{\partial v}{\partial W} = 0. \tag{1.112}$$

The system (1.112) has an integral

$$(v+\theta)^2 = W. \tag{1.113}$$

The solution of eqn (1.112) may be written as

$$v = C - \frac{2\theta}{3}, \quad W = \left(C + \frac{\theta}{3}\right)^2, \tag{1.114}$$

where C is an arbitrary constant. Omitting the function θ from eqn (1.114), we find the travelling wave, eqn (1.107) in the form

$$\mathrm{W} = \tfrac{1}{4}(3C - v)^2. \tag{1.115}$$

Equation (1.115) describes a family of explicit solutions, depending on the parameter C. Being similar to nonstationary Riemann waves in one-dimensional hydrodynamics, these solutions describe the spreading of the pulse, resulting in the formation of 'spillovers' at the pulse periphery. The dynamics of these processes depends on the initial profile of the normalized frequency envelope $v_0, (q)$. Using an inverse dependence $q_0 = q_0(v)$ and combining it with eqns (1.114) and (1.111), we can determine the function $P(\theta)$. Thus, in the case of a 'saturated' profile of initial frequency modulation,

$$v_0(q) = \frac{aq^2}{1+q^2}, \tag{1.116}$$

where a is an arbitrary constant, we obtain

$$P(\theta) = \left[\frac{3C - 2\theta}{3(a-C) + 2\theta}\right]^{1/2}. \tag{1.117}$$

The dependence $q = q(\theta)$, eqn (1.110), may now be considered as a cubic equation with respect to the quantity θ, containing τ as a parameter. This dependence, being single-valued at the plane $\tau = 0$, becomes triple-valued starting with some value $\tau = \tau_c$.

Such a hysteresis effect results in the formation of a 'spillover' due to pulse spreading and an increase in its slope. The method of analysis of such

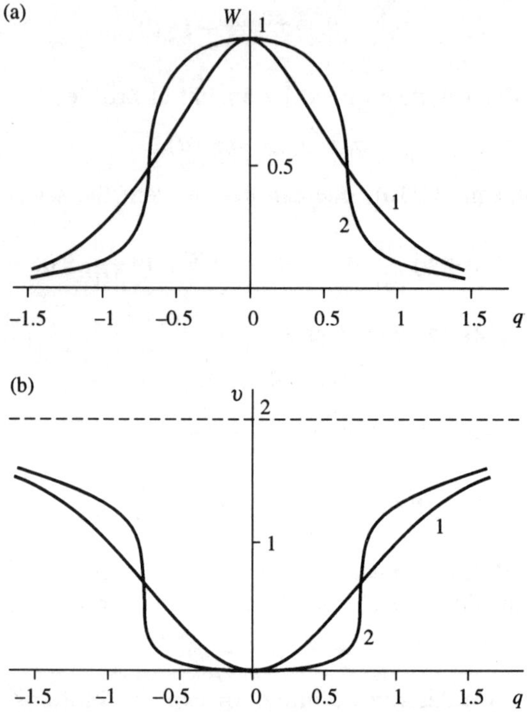

Fig. 1.14 Nonlinear evolution of the running wave (eqn 1.115) $(C=2\cdot 3^{-1})$ without a pedestal, $W(q \to \pm\infty)=0$. The intensity profiles W and spectral envelopes υ are plotted against travelling coordinate q in (a) and (b) respectively. Curves 1 relate to the initial profiles $W_0(q)$ (eqn 1.119) and $\upsilon_0(q)$ (eqn 1.116) $(a=2)$; curves 2 show the self-steepening of the pulse fronts and self-demodulation of its central part at the point $\tau=0.5$. Relatively small peak perturbations W = 0.8, and V = 0.2 are not shown on curves 2.

phenomena is well known, for example from the theory of a nonlinear pendulum in mechanics or from the theory of optical bistability (Gibbs 1985). To find the coordinate and time of appearance of the spillover, one must solve eqn (1.110) together with an equation describing the appearance of a vertical tangent to the intensity profile $W(\tau, q)$. This equation, derived from eqn (1.117), is

$$\left.\frac{\partial q}{\partial \theta}\right|_{\tau} = 0. \qquad (1.118)$$

Thus, considering for example the values of the constants $a=2$, $C=2/3$, relevant to the initial intensity profile eqn (1.115), (Fig. 1.14a),

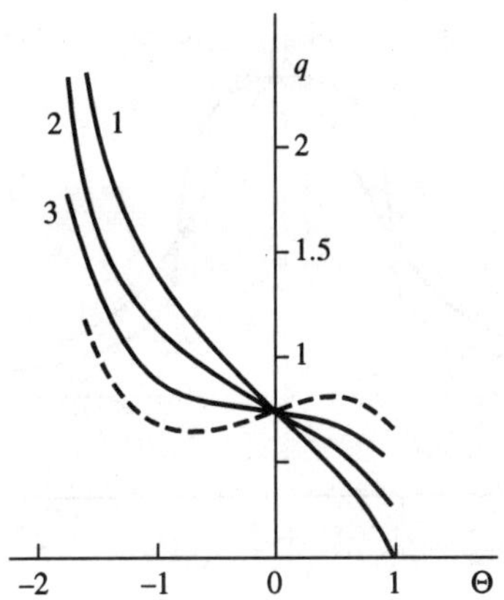

Fig. 1.15 Formation of a triple-valued dependence of travelling coordinate q on the power parameter θ (eqn 1.110) in the course of propagation of the running wave shown in Fig. 1.14. Curves 1, 2 and 3 relate to normalized distances $\tau = 0$; $\tau = 0.25$, and $\tau = 0.5$ respectively ($\tau < \tau_{cr} = 0.51$); the broken line illustrates the case $\tau > \tau_{cr}(\tau = 0.6)$.

$$W_0 = \frac{1}{(1+q^2)^2}, \tag{1.119}$$

we obtain eqn (1.110) in the form

$$q = \theta\tau + \left(\frac{1-\theta}{2+\theta}\right)^{1/2} \tag{1.120}$$

The values of the dimensionless parameter θ are located in the interval $-2 \leqslant \theta \leqslant 1$. The graph of eqn (1.120) corresponding to different values of the parameter τ, illustrates the occurrence of an N-type hysteresis curve when $\tau > \tau_c = 0.51$ (Fig.1.14). The steepening of the intensity and spectral envelopes, causing flattening of the pulse peak and spreading of the domain of pulse demodulation ($v \to 0$), is shown in Fig. 1.14 at the point $\tau = 0.5$, close to the critical point τ_c.

By varying the constant C and initial profile $v_0(q)$ one can describe a family of travelling modulated waves, eqn (1.115), with different values of modulation index M and minimum intensity W_{min}; the greatest intensity is W_{max}. M is defined as

$$M = \frac{W_{max} - W_{min}}{W_{max} + W_{min}} \tag{1.121}$$

Thus the envelope (1.119) relates to the value $M = 1$; considering for example the initial distribution $v_0(q)$ in the form (1.116) and choosing the

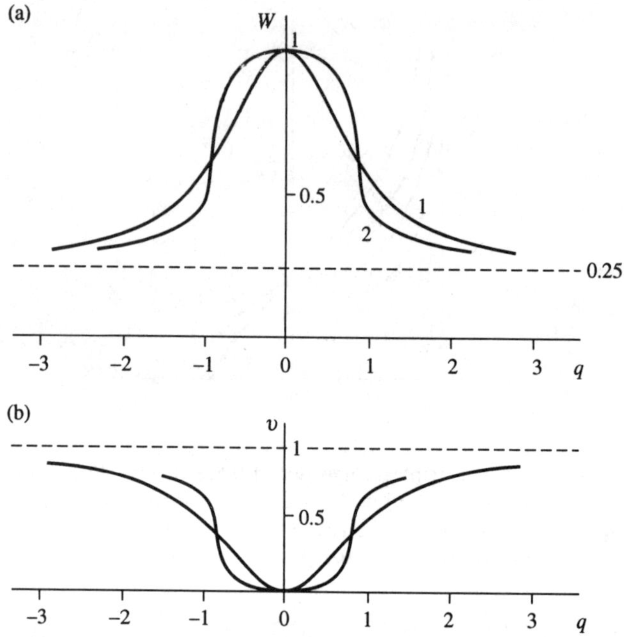

Fig. 1.16 (a) Self-steepening of the fronts and (b) self-demodulation of the central part of the running wave (eqn 1.115) ($C = 2 \cdot 3^{-1}$) with nonzero pedestal $W(q \to \pm\infty) = 0.25$. Curves 1 relate to the initial intensity W_0 (eqn 1.122) and frequency v_0 (eqn 1.116) ($a = 1$) envelopes; Curves 2 show the reshaping of the intensity profile W and the frequency 'chirp' at the point $\tau = 0.45$; relatively small peak perturbations $W_{\max} = 0.75$, $V_{\max} = 0.27$ are not shown on curves 2.

the values $C = 2/3$, $a = 1$, we obtain the travelling pulse (1.115) with the initial intensity profile

$$W_0(q) = \frac{1}{4}\left(2 - \frac{q^2}{1 + q^2}\right)^2. \tag{1.122}$$

This pulse, characterized by modulation index $M = 0.25$, is depicted in Fig. 1.16. The spreading of the pulse with nonzero pedestal ($M < 1$) shown in Fig. 1.16 possesses the same tendency to steepening of the front as a pulse without pedestal (Fig. 1.14).

It is important to emphasize that the modulation index M, eqn (1.121), as well as the maximum and minimum values of the intensity profile, are changing in the course of nonlinear evolution of the pulses, eqn (1.107). It is of interest to mention that the nonlinear dynamics of pulses with $M < 1$ cannot be treated analytically by the interchange of independent variables and functions used in Subsection 1.3.1.

Hence the possibilities of using nonlinear, nonstationary geometrical optics for the development of power-controlled spectral tuning of solitary pulses are characterized by the following features:

1. The coupled evolution of intensity and spectral profiles of powerful pulses provides controlled shaping of the time-dependent frequency envelope at a given point within the dispersive medium. Thus, considering the propagation of a bell-like picosecond pulse, eqn (1.92) ($\lambda = 1.5\,\mu$m, $T = 2$ ps, the peak intensity 100 W, $|\chi_{01}| = 100$) along a silica fibre with a core diameter of 4 μm and a dispersive length, eqn (1.59), $L_\omega = 200$ m and using the value $|v|_{\max} = 0.9$, as shown in Fig. 1.12b, one can evaluate the maximum spectral broadening $\Delta\beta \approx 80\,\text{cm}^{-1}$. This spectral broadening is accumulated along the path $\tau = 0.5$ (Fig. 1.12b); according to eqn (1.68) this normalized value τ relates to a length of fibre of ~8 m.

2. The strong influence of initial frequency modulation of the pulse $v_0(q)$ on its nonlinear evolution may be visualized by comparison of the dynamics of an initially unmodulated (Fig. 1.13) and modulated (Fig. 1.14) bell-like envelope, travelling in glass fibre ($\chi > 0$) in the range of normal dispersion ($v_\omega > 0$). The peak of the modulated pulse (Fig. 1.13) decreases in the course of evolution.

3. Nonlinear geometric optics provides a unified and standardized method for analysis of coupled power-dependent deformations of intensity profiles and spectral bandwidths of ultrashort pulses in dispersive media. The flexibility of deformation regimes controlled by the initial envelope favours the adjustment of pulse parameters at the output of the nonlinear system.

This analysis of large-scale time-dependent optical phenomena is applicable when the final values of the radiation wavelengths have no influence on the effect discussed. Small-scale wavelength-dependent nonstationary optical effects giving rise to profound modulation of radiation fluxes are considered below.

1.4 Generation of a periodic train of broadband pulses owing to modulational instability in optical fibres

The power-dependent properties of dispersive media may be used both for tunable reshaping of solitary pulses as discussed above and for the formation of periodic trains of such pulses due to self-lamination of initially continuous radiation fluxes. Such self-lamination can arise in the development of modulational instability in nonlinear dispersive systems in the case $\chi v_\omega < 0$ (Whitham 1967) and in particular in optical fibres in the IR range ($\chi > 0$, $v_\omega < 0$) (Hasegawa 1984). This effect, confirmed experimentally by Tai *et al.* (1986a), proves to have potential for the design of pulse-series generators with high repetition rate for communication systems and for spectral transformation of optical signals. The dynamics of modulational instability, depending on the finite values of the optical wavelengths, is described by the periodic solutions of the nonlinear Schrödinger equation (1.57).

The simplest examples of such solutions are associated with stationary periodic profiles of intensity and modulation, described by eqns (1.67) and, (1.70) in the limit $\partial/\partial\eta \to 0$ (Usunov 1990). The correlation between the distributions of intensity $W(q)$ and modulation $v(q)$, habitual to these distributions, is obtained from eqn (1.71):

$$v(q) = \frac{v_0}{W(q)}. \tag{1.123}$$

Here v is the smallest value of the 'chirp' parameter. Substituting eqn (1.123) in eqn (1.67), one can derive the periodic profiles $W(q)$ and $v(q)$, described by Jacobian elliptic functions. The envelopes and periods of these profiles depend essentially upon the modulation parameter v_0.

Unlike such stationary propagation of waves with permanent amplitude–chirp distributions, the consideration of nonstationary pulse regimes by means of the general equation (1.57) reveals the influence of modulational instability on the time-dependent lamination of a wave beam, trapped in a nonlinear waveguide. The first stage of this instability, promoting the growth of harmonic pepturbations on the background of the initial distribution of field structure, can be analysed through linearization of the master equation (1.57). To find the increments of these processes, it is convenient to renormalize the variables in eqn (1.57), introducing instead of an arbitrary duration of the pulse T, eqn (1.58), the natural time scale T_0 associated with the dispersion of the medium and the free-space radiation wavelength λ_0 (Tracy *et al.* 1984):

$$T_0 = \frac{\sqrt{-\lambda_0 v_\omega}}{v_g\sqrt{n_0}}. \tag{1.124}$$

Rewriting eqn (1.57) with new dimensionless variables x, y and the normalized function U,

$$x = \frac{2\pi n_2 |E_0|^2 z}{\lambda_0}, \quad y = \frac{(t - z v_g^{-1})\sqrt{(2\pi n_2 |E_0|^2)}}{T_0}, \quad U = F\exp(-\mathrm{i}x), \tag{1.125}$$

one can obtain the equation governing the function U in the form

$$\mathrm{i}\frac{\partial U}{\partial x} + \frac{1}{2}\frac{\partial^2 U}{\partial y^2} - U + |U|^2 U = 0. \tag{1.126}$$

Let us consider, using eqn (1.125), the stability of a continuous flux of radiation with constant amplitude $|U|$, equal to unity: $U = \exp(\mathrm{i}\varphi)$. Expanding this solution near the point $(U = 1, \varphi = 0)$, one can obtain

$$U = 1 + \left[a\left(1 + \frac{i\delta}{k^2}\right) e^{\delta x} + b\left(1 - \frac{i\delta}{k^2}\right) e^{-\delta x} \right] \cos(ky). \tag{1.127}$$

Here a and b are independent small constants ($|a| \ll 1$, $|b| \ll 1$); the increment of instability is $\delta = k\sqrt{(1 - 0.25k^2)}$. This increment, being real in the range $-2 \leqslant k \leqslant 2$, reaches its maximum value $\delta = 1$ when $k = \sqrt{2}$.

However, this linearized approximation is valid only at the beginning of the evolution. To analyse the development of modulation instability in the whole time domain (formally $-\infty < q < \infty$), we have to find some further exact analytical nonstationary solutions of eqn (1.26). Since the value $k = \sqrt{2}$ relates to the largest increment of instability, this case is examined here.

1.4.1 Time-dependent self-lamination of a powerful radiation flux

A rigorous general theory of asymptotic behaviour of the fields described by the nonlinear Schrödinger equation (1.57) was advanced by Zakharov *et al.* (1980) in the framework of the so-called 'inverse scattering method'. This approach has been abandoned in mathematics; moreover, it proves to be difficult by this asymptotic method to obtain the amplitude and spectral envelopes of the fields, formed during a finite time of evolution. To simplify the problem without loss of generality, a suitable analytical method operating outside the scope of perturbation theory is considered here. This method contains the following procedures.

1. The function U, determined by eqn (1.126), is presented in the form

$$U = P + iQ. \tag{1.128}$$

The real functions P and Q are determined by the pair of equations obtained through substitution of eqn (1.128) in eqn (1.126):

$$\left.\begin{aligned} \frac{\partial P}{\partial x} - Q + \frac{1}{2}\frac{\partial^2 Q}{\partial y^2} + (P^2 + Q^2)Q = 0, \\ -\frac{\partial Q}{\partial x} - P + \frac{1}{2}\frac{\partial^2 P}{\partial y^2} + (P^2 + Q^2)P = 0. \end{aligned}\right\} \tag{1.129}$$

2. Let us suppose the functions P and Q to be linked by a linear correlation containing the auxiliary functions $\gamma(x)$ and $\mu(x)$. (Akchmediev *et al.* 1988):

$$P = \gamma(x)Q + \mu(x). \tag{1.130}$$

The aim of this representation is to reduce the system of partial differential equations (1.129), determining two functions P and Q, to one ordinary differential equation for one function (P or Q). By choosing the suitable functions $\gamma(x)$ and $\mu(x)$, one can perform this reduction. Thus, if

$$\gamma(x) = \mu(x) = -\operatorname{th} x, \tag{1.131}$$

we obtain the Bernoulli equation governing the function Q:

$$\frac{\partial Q}{\partial x} + Q^2 \operatorname{th} x + Q \operatorname{th} x = 0. \tag{1.132}$$

The general solution of eqn (1.132) is

$$Q = \frac{C(y)}{\operatorname{ch} x - C(y)}. \tag{1.133}$$

Here the constant of integration with respect to x may depend on the variable y.

3. To find the function $C(y)$ in eqn (1.133) let us substitute eqn (1.133) in each of the equations obtained previously through substitution of eqn (1.130) in the system (1.129). Both of these operations lead to the same equation:

$$\frac{\partial^2 C}{\partial y^2} + \frac{2}{\operatorname{ch} x - C}\left(\frac{\partial C}{\partial y}\right)^2 + \frac{2(C^2 + C \operatorname{ch} x - 1)}{\operatorname{ch} x - C} = 0. \tag{1.134}$$

The solution of nonlinear equation (1.134) is

$$C(y) = \frac{1}{\sqrt{2}} \cos(\sqrt{2}\, y). \tag{1.135}$$

Finally, combining eqns (1.130), (1.131), (1.133) and (1.135), we obtain the exact nonstationary solution of eqn (1.126) in a complex form:

$$U(x, y) = \frac{-\sqrt{2} \operatorname{sh} x + \mathrm{i} \cos(\sqrt{2} y)}{\sqrt{2} \operatorname{ch} x - \cos(\sqrt{2} y)}. \tag{1.136}$$

The envelope (1.136) is one of the simplest nonstationary periodic solutions of the nonlinear Schrödinger equation. In the limit $x \to -\infty$, this solution coincides with the expansion (1.127), the constant b being zero. Thus this solution describes the development of modulational instability of an initially unmodulated field with amplitude ($U|_{x \to -\infty} = 1$).

The time-dependent self-lamination of this field and its periodic spectral broadening are shown in Fig. 1.17. Representing the function U, eqn (1.136) in the form $U = A \exp(\mathrm{i}\varphi)$, one can see that the modulation amplitude A increases in the course of evolution, achieving its largest value at the point $x = 0$ (Fig. 1.17a); the radiation flux proves to be totally stratified at this point. The modulation weakens in the domain $x > 0$ [$A(x) = A(-x)$], tending to zero when $x \to \infty$. The distribution of the spectral parameter $v = \partial\varphi/\partial y$ within each period of modulation proves also to be periodic and antisymmetric with respect to y: $v(y) = -v(-y)$ (Fig. 1.17b). This frequency chirp vanishes when $y \to \pm\infty$.

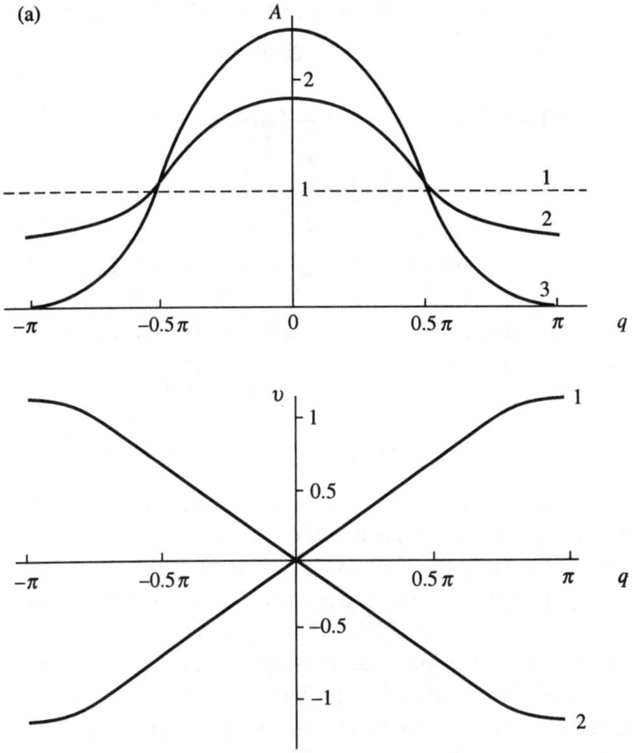

Fig. 1.17 (a) Formation of a solitary pulse through development of modulational instability of an initially constant radiation flux; (b) and oscillations of its spectral envelopes, described by eqn (1.136). The values $-\pi \leqslant q \leqslant \pi$ relate to one period of oscillation. Curves 1, 2, and 3 in (a) show the amplitude profiles A at the points $\eta = \pm\infty$, $\eta = \pm 0.5$, and $\eta = 0$ respectively. Curves 1 and 2 in (b) correspond to the spectral envelopes v at the points $\eta = 0.5$ and $\eta = -0.5$ respectively.

Considering the nonstationary periodic solutions, it is convenient to analyse their coordinate-dependent spectra. Representing the solution (1.136) by the Fourier expansion

$$U(x, y) = f_0(x) + 2\sum_{n=1}^{\infty} f_n(x) \cos(n\sqrt{2}\,y), \tag{1.137}$$

and using the formulae for Fourier coefficients $f_n(x)$, we can find these coefficients analytically:

$$\left.\begin{aligned} f_0(x) &= \frac{\mathrm{i}\sqrt{2}\,\mathrm{sh}\,x + \sqrt{2}\,\mathrm{ch}\,x - \sqrt{2\,\mathrm{ch}^2 x - 1}}{\sqrt{2\,\mathrm{ch}^2 - 1}} \\ f_n(x) &= \frac{\sqrt{2}(\mathrm{ch}\,x + \mathrm{i}\,\mathrm{sh}\,x) + (\sqrt{2}\,\mathrm{ch}\,x - \sqrt{2\,\mathrm{ch}^2 x - 1})^n}{\sqrt{2\,\mathrm{ch}^2 - 1}} \end{aligned}\right\} \quad (1.138)$$

These coefficients obey the condition

$$|f_0(x)|^2 + 2\sum_{n=1}^{\infty} |f_n(x)|^2 = 1. \quad (1.139)$$

The spectral parameters $|f_n(x)|^2$ possess the following properties:

$$\left.\begin{aligned} &\lim |f_0(x)|_{x \to \pm\infty} = 1, \quad \lim |f_n(x)|_{\substack{x\to\pm\infty \\ n\neq 0}} = 0, \\ &|f_0(x)|^2_{x=0} = (\sqrt{2} - 1)^2, \quad |f_n(x)|^2_{x=0} = \sqrt{2}\,(\sqrt{2} - 1)^{2n} \end{aligned}\right\} \quad (1.140)$$

Thus, in the course of evolution from $x = -\infty$ to $x = 0$, all the spectral components increase; at the point $x = 0$ the maximim power is contained in the first side lobe ($n = 1$). The initial state of the field will be recovered through spectral demodulation during the subsequent evolution from $x = 0$ to $x \to \infty$.

The analysis of the amplitudes and spectra of modulational instability shows that the initial growth of field perturbations changes after some time to their decline and a return to the initial state of the field.

This example relates to the case of an initially unmodulated field. Proceeding in a similar fashion, one can analyse the amplification of initial modulation in the course of development of modulational instability. Such amplification results in a rapid generation of a periodic train of solitary pulses. To provide a flexible representation of initial modulation it proves to be convenient to use the Jacobian elliptic functions $\mathrm{sn}(x,p)$, $\mathrm{cn}(x,p)$ and $\mathrm{dn}(y,p)$, depending on the coordinate x, eqn (1.125), and modulus p; these functions are connected by the formulae (Janke *et al.* 1960):

$$\mathrm{cn}^2(x,p) = 1 - \mathrm{sn}^2(x,p), \quad \mathrm{dn}^2(x,p) = 1 - p^2 sn^2(x,p). \quad (1.141)$$

Employing eqns (1.141) and restricting ourselves for simplicity to the case $0 \leqslant p \leqslant 1$, we can choose the auxiliary functions $\gamma(x)$ and $\mu(x)$, eqn (1.131) in a form suitable for the problem discussed (Akchmediev and Korneev 1987):

$$\gamma(x) = -\frac{p \cdot \mathrm{sn}(x/p,p) \cdot \mathrm{cn}(x/p,p)}{\mathrm{dn}(x/p,p)}, \quad \mu(x) = -\mathrm{sn}(x/p,p). \quad (1.142)$$

Making use of eqns (1.142), (1.130), and (1.129) and omitting the intermediate steps, we obtain the family of nonstationary solutions of eqn (1.126):

$$\left.\begin{aligned} U &= \frac{-\mathrm{sn}(x/p,p) + \mathrm{i}A(y)\,\mathrm{dn}(x/p,p)}{1 - p\cdot A(y)\,\mathrm{cn}(x/p,p)}, \\ A(y) &= \mathrm{cn}\left[\sqrt{(2/p)}y; \sqrt{\left(\frac{1-p}{2p(1+p)}\right)}\right]. \end{aligned}\right\} \tag{1.143}$$

This solution can be checked directly by substitution of eqns (1.143) in eqn (1.126).

The time-dependent nonsinusoidal filamentation of the field structure described by formulae (1.143) is shown in Fig. 1.18. Regarding these graphs, it should be mentioned that the expressions (1.143) form a family of solutions depending on the modulus p of the elliptic functions. This modulus, determining the shape of nonsinusoidal oscillations, is a free parameter for these solutions. The solutions (1.143) are valid in the range $0 \leqslant p \leqslant 1$ (Fig. 1.18a and b); to consider larger values of this free parameter ($1 < p < \infty$) we can use some formulae from the theory of elliptic functions, e.g. (Janke *et al.* 1960)

$$\mathrm{sn}\left(pu, \frac{1}{p}\right) = p\cdot \mathrm{sn}(u,p), \quad \mathrm{cn}\left(pu, \frac{1}{p}\right) = \mathrm{dn}(u,p), \tag{1.144}$$

Using these transformations and introducing the parameter $\kappa = p^{-1}$, we can rewrite the solutions (1.143) in a form valid in the range $1 \leqslant p < \infty$ ($0 \leqslant \kappa \leqslant 1$):

$$\left.\begin{aligned} U &= \frac{-\kappa\cdot \mathrm{sn}(x,\kappa) + \mathrm{i}B(y)\,\mathrm{cn}(x,\kappa)}{1 - \kappa^{-1}B(y)\,\mathrm{dn}(x,\kappa)}, \\ B(y) &= \frac{\kappa}{\sqrt{1+\kappa}}\,\frac{\mathrm{cn}(\sqrt{(1+\kappa)}y; \sqrt{(1-\kappa)(1+\kappa)^{-1}}}{\mathrm{dn}(\sqrt{(1+\kappa)}y; \sqrt{(1-\kappa)(1+\kappa)^{-1}}}. \end{aligned}\right\} \tag{1.145}$$

This solution, illustrating the nonlinear anharmonic oscillations of the field's envelope, is shown in Fig. 1.18c. In the limiting case $p = 1$, the solution (1.143) can be reduced to the envelope (1.136) illustrating the growth of sinusoidal perturbations of initially constant amplitude.

The dashed lines in Fig. 1.18 show the initial amplitude profiles at the input of the fibre ($x_0 = -p\cdot \mathcal{K}(p)$); these profiles at the point $x = 0$ are depicted by the solid lines. Comparing these lines, one can see that the development of modulational instability transforms the initially modulated signals with modulation index $M < 1$ to solitary pulses without pedestal ($M = 1$). This process may be of interest for the nonlinear conversion of amplitude-modulated radiation admitted to the fibre at the point x_0 to a periodic train of solitary pulses emerging from the fibre span at the point

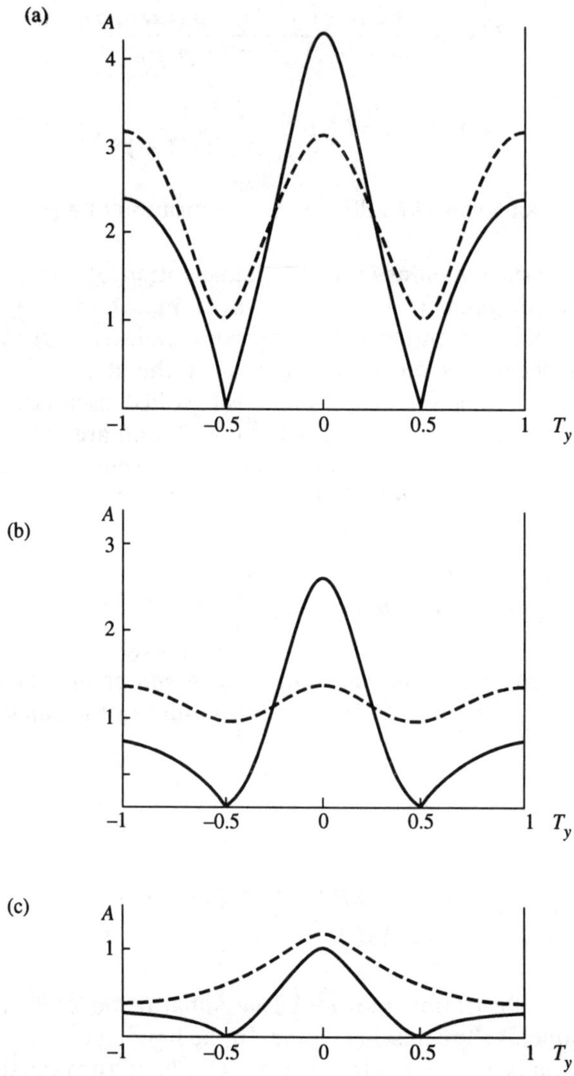

Fig. 1.18 Formation of a solitary pulse (solid lines) due to nonlinear oscillations of initial periodic amplitude envelopes (broken lines). The amplitude profiles A relate to one temporal period T_y ($-1 \leqslant T_y \leqslant 1$) of nonsinusoidal oscillation (eqn 1.147). The distance between the positions shown by the broken and solid lines is equal to one-quarter of the period of spatial nonsinusoidal oscillations T_x. (a), (b) and (c) correspond to values of the modulus of the Jacobian functions $p = 0.1$, 0.5 (eqn 1.143), and $\kappa = 0.5$ (eqn 1.145) respectively.

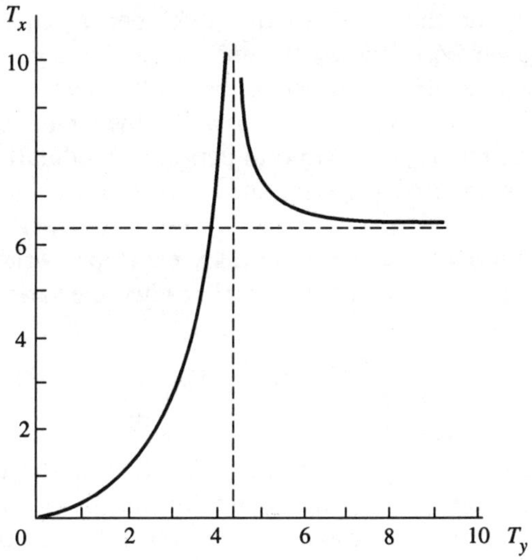

Fig. 1.19 Dispersive curve relating the spatial T_x and temporal T_y periods of non-linear anharmonic oscillations, eqns (1.146), (1.147). The vertical and horizontal broken lines relate to values of modulus of the Jacobian functions $p = 1$ ($T_y = \sqrt{2}\pi$, $T_x \to \infty$) and $\kappa = 0$ ($T_y \to \infty$, $T_x = 2\pi$) respectively.

$x = 0$. The dimensionless length of this span L most be equal to one quarter of the spatial period T_x. The period T_x is determined by complete elliptic integrals of the first kind:

$$\left.\begin{aligned} T_x &= 4p\mathcal{K}(p), \quad 0 \leqslant p \leqslant 1, \\ T_x &= 4\mathcal{K}(\kappa), \quad 0 \leqslant \kappa \leqslant 1. \end{aligned}\right\} \tag{1.146}$$

The normalized temporal period T_y is

$$\left.\begin{aligned} T_y &= 2\sqrt{2p}\,\mathcal{K}\left(\sqrt{\frac{1-p}{2}}\right), \quad 0 \leqslant p \leqslant 1, \\ T_y &= \frac{4}{\sqrt{1+\kappa}}\mathcal{K}\left(\sqrt{\frac{1-\kappa}{1+k}}\right), \quad 0 \leqslant \kappa \leqslant 1. \end{aligned}\right\} \tag{1.147}$$

Thus the spatial and temporal periods of these oscillations depend upon the waveform, determined by parameter p. The graph $T_x = T_x(T_y)$ may be considered by analogy with the dispersive curve well known in the theory of harmonic oscillations. The dependence of T_x on T_y is shown in Fig. 1.19.

By selecting the shape of the output signal and fixing the modulation period, one can find the length of the fibre span L, eqn (1.146) and the average peak power W_0 of the input signal which are required for the transformation discussed. Thus, for an optical fibre ($n_0 = 1.5$; $\lambda = 1.55\,\mu$m; $v_\omega = -8 \times 10^{-10}$ m; $n_2 = 1.2 \times 10^{-22}\,\mathrm{m^2\,V^{-2}}$) the value of the modulus $k = 2$ (Fig. 1.18c) relates to a temporal period of modulation of ~6 ps and a length of fibre span of ~100 m; the average power W_0 in this case is $W_0 = 1.5$ W.

The stratification of the pulse complex envelope, written in the form $U(x, y) = A \exp(\mathrm{i}\varphi)$, is accompanied by considerable variations of its spectral content:

$$\Delta\omega = -\frac{\sqrt{2\pi n_2 |E_0|^2}}{T_0}\frac{\partial\varphi}{\partial y} \tag{1.148}$$

In a simple case of evolution of the field (1.136), the maximum value $|\Delta\omega|_0$ for each point x may be found by means of values $v = \partial\varphi/\partial y$, depicted in Fig. 1.17b. Thus, using the value $v = 1.15$ ($x = 0.5$) and supposing $n_2|E_0|^2 = 2 \times 10^{-8} n_0$, $T_0 = 1.5 \times 10^{-16}$ s, eqn (1.124), one can estimate $|\Delta\omega|_0 = 0.5$ THz.

This method of shaping trains of picosecond pulses in fibre circuits proves to be very sensitive to the modulation of the input continuous laser beam. Flexibility of this shaping may be provided by interference of two fields with close frequencies, radiated by two sources. Such an approach is promising for the generation of picosecond pulse series with high repetition rate ν_0 (e.g. $\nu_0 = 0.3$ THz, the pulse halfwidth being ~0.5 ps: Tai *et al.* 1986b) and for ultrafast switching of signals in fibre interferometers (Islam *et al.* 1988). Moreover, these pulse trains may be useful for the design of high-bit-rate all-optical information systems.

1.4.2 Modulational instability of radiation in periodic waveguide structures

The above-mentioned possibilities for the generation of pulse trains through modulational instability of a wave in a nonlinear dispersive medium may be diversified by periodical spatial variation of the refractive index of this medium in the direction of wave propagation. These phenomena are of interest for wave propagation in modulated waveguides. Such waveguide modulation through periodic variation of the radius of the core $a = a_0 + a_1 \cos(k_0 z)$ may be produced by changes in the speed of fibre-drawing during fabrication. Another method of waveguide modulation is based not on mechanical profiling of the thickness but on periodic variation of the refractive index of the core:

$$n(\omega, z) = n_0(\omega) + \Delta n \cos(k_0 z). \tag{1.149}$$

Parameter k_0 in eqn (1.149) is related to the spatial period of modulation Λ: $k_0 = 2\pi\Lambda^{-1}$. Thus GeO_2-doped silica fibres are known to possess a photosensitivity, that permits the use of holography to create of phase gratings which can be embedded within the cores of fibres. Since such a waveguide can be considered as a Bragg filter, its properties are highly dependent on the radiation wavelength (Winful 1985).

To reveal the features of the development of modulational instability in a modulated waveguide, one can represent the power-independent part $n_0(\omega, z)$ of the nonlinear refractive index, eqn (1.43), in the form (1.149). Supposing the refractive index (1.149) to be a slowly varying function of coordinate z at the distance λ_0 (the free-space wavelength of the trapped radiation), one can derive the nonlinear Schrödinger equation, generalizing eqn (1.126) for the periodic structure:

$$\left.\begin{aligned} &\mathrm{i}\frac{\partial U}{\partial x} + \frac{1}{2}\frac{\partial^2 U}{\partial y^2} + [1 + \alpha\cos(k_1 x)](-1 + |U|^2)U = 0, \\ &k_1 = \frac{k_0}{2\pi n_2 |E_0|^2}, \quad \alpha = \frac{\Delta n}{n_0}. \end{aligned}\right\} \tag{1.150}$$

The values of the variables x, y and function U are given by eqns (1.125).

Let us analyse in the framework of eqn (1.150) the stability of propagation of a continuous wave beam with a constant amplitude equal to unity $U = \exp(\mathrm{i}\varphi)$, through the periodic structure, eqn (1.149). This wave beam was shown to be unstable, eqn (1.127), with respect to its periodical perturbations in the case of vanishing spatial modulation, $\alpha \to 0$, reducing eqn (1.150) to eqn (1.126). Proceeding in a similar fashion, let us examine the stability of the perturbed wave distribution

$$U = [1 + U(x)\cos(by)]\exp(\mathrm{i}\varphi). \tag{1.151}$$

Here $U = P + \mathrm{i}Q$, P and Q are the real functions, and $|U|^2 \ll 1$. Substituting eqn (1.151) in eqn (1.150), one can obtain the Mathieu equation for the function $P(x)$:

$$\frac{\partial^2 P}{\partial x^2} + \left[b^2\left(\frac{b^2}{4} - 1\right) - \frac{b^2\alpha}{2}\cos(k_1 x)\right]P = 0. \tag{1.152}$$

The function $Q(x)$ obeys the same equation.

By fixing the parameter q characterizing the depth of modulation of the refractive index, eqn (1.150), one can find an unlimited sequence of values of wavenumber b, relevant to the periodic solutions of eqn (1.152); these solutions are known to be the real periodic Mathieu functions with a period equal to 2π (Janke *et al.* 1960). The multitude of other values of wavenumbers b which do not belong to this sequence relate to the unstable regimes of propagation of the field, eqn (1.151).

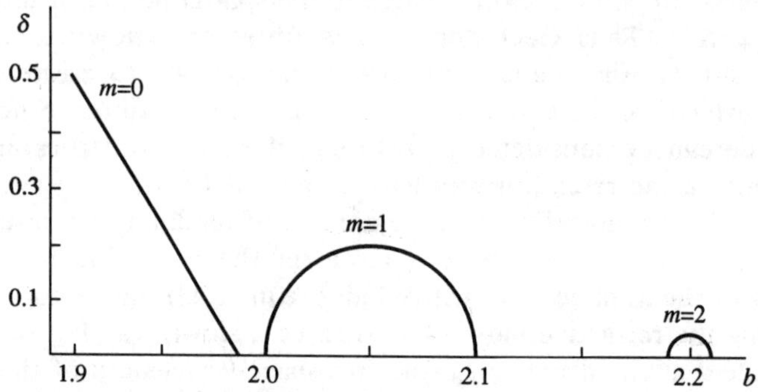

Fig. 1.20 Zones of modulational instability for the linearized equation with periodic coefficients (eqn 1.152) on a plane of normalized varlables (δ, b); the instability increment δ is plotted vs. the perturbation wavenumber b; $\alpha = 0.1$, $k_1 = 1$, m is the number of the instability zone.

An important feature of this instability, distinguishing the unstable regimes in homogeneous and periodic media, is associated with the zones of unstable values of perturbation wavenumbers: perturbations in a homogeneous medium, eqn (1.127), were shown to increase in only one domain on the (δ, k) plane $(-\sqrt{2} < k < \sqrt{2})$, whereas periodic nonlinear media are characterized by a system of zones of instability on the plane (δ, b). The scales and arrangement of these zones depend on the values α and b, eqn (1.151); some such areas are depicted in Fig. 1.20. These new zones may be useful for controlled generation of pulse trains in periodic nonlinear optical circuits and crystals (Feng and Kneubuhl 1993).

1.5 Diffraction of polychromatic wave beams and short pulses

The investigation of the diffraction of long trains of quasi-monochromatic waves by various conducting and dielectric targets, layered bodies, and optical and quantum objects was acted for many decades as a catalyst for the development of a multitude of classical concepts in wave theory and the relevant mathematics (Morse and Feshbach 1953). These traditional concepts are now widely used in computational electromagnetics, with adaptation of such idealized models to the analysis of real scatterers located in real media. The new trends in this field are connected wlth pulse optics and recently posed problems of diffraction and scattering of ultrashort wideband pulses.

To illustrate the specificity of these problems, it is worth reconsidering some of the simplest topics (e.g. Fraunhofer diffraction at slits and circular openings or scattering of a plane wave on a sphere), indicating the distinctions associated with the finite radiation bandwidth. Let us start from the expression for the field $E(x, y)$ at a distance z from a planar distribution $\Psi(\varepsilon, \eta)$, the observation plane (x, y) and the aperture plane (ε, η) being parallel (Born and Wolf 1964):

$$E = \frac{k}{2\pi} \iint \frac{\Psi(\varepsilon, \eta)}{R} \exp(ikR) \, d\eta \, d\varepsilon, \tag{1.153}$$

$$R = \sqrt{[z^2 + (x-\varepsilon)^2 + (y-\eta)^2]}. \tag{1.154}$$

Using the Huygens–Fresnel approximation $z^2 \gg (x-\varepsilon)^2 + (y-\eta)^2$, one can rewrite eqn (1.153) as

$$E(\omega) = \frac{ik}{2\eta z} \exp(ikz) \iint d\varepsilon \, d\eta \, \Psi(\varepsilon, \eta) \exp\left\{\frac{ik[(x-\varepsilon)^2 + (y-\eta)^2]}{2z}\right\}. \tag{1.155}$$

This equation will be used below for both 1D and 2D distributions of the illuminated area $\Psi(\varepsilon, \eta)$.

Some numerical results concerning the wideband diffraction phenomena can be considered through formal summation of monochromatic effects associated with the cumbersome integration in the complex ω-plane; an example of such approach will be discussed in Section 1.8. This section on the other hand is devoted to physically meaningful effects of the final bandwidth of wave fields passing through apertures.

1.5.1 Diffractive variations of the frequency spectrum of a broadband source

Propagation of a polychromatic radiation beam through an aperture can produce spatial redistribution of frequencies in the diffracted beam (Wolf 1986). To analyse this effect, let us consider the spectral intensity of the axisymmetrical beam $I_\omega(\rho, z)$ at a distance z from the aperture. The intensity $I_\omega(\rho, z)$ may be presented in the form

$$I_\omega(\rho, z) = G_0(\omega) M_\omega(\rho, z). \tag{1.156}$$

Here $G_\omega(\omega)$ is the spectrum of the source, and the spectral transfer function $M_\omega(\rho, z)$ contains all the information concerning the diffractive variations of this spectrum; let us suppose the function $M_\omega(\rho, z)$ to be independent of the frequency at the aperture plane: $M_\omega(\rho, 0) = M(\rho)$. The emergence of a dependence of M_ω on the frequency ω in the course of propagation $(z > 0)$ will indicate that the beam's spectrum is dependent on the distance z.

Restricting ourselves for simplicity for a spatially coherent source and the paraxial approximation, we can write the function M_ω, eqn (1.156), as

$$M_\omega(\rho, z) = |E(\omega)|^2. \tag{1.157}$$

Here $E(\omega)$ is given by eqn (1.155). By calculating the diffractive fields $E(\omega)$ for some field distribution at the aperture $\Psi(\varepsilon, \eta)$, one can analyse the spatial structure of the diffractive pattern of the wideband beam in both nondispersive ($k = \omega c^{-1}$) and dispersive ($k = \omega n(\omega) c^{-1}$) media. It is useful to discuss here the coherent Gaussian distribution

$$\Psi|_{z=0} = \exp\left(-\frac{\rho^2}{2\rho_0^2}\right). \tag{1.158}$$

The properties of series of elements of integrated and waveguide optics result in the formation of Gaussian amplitude–phase distributions of light waves. Such distributions are formed as the eigenmodes of both open laser cavities with spherical mirrors and parabolic-index waveguides. Moreover, the Gaussian beams are known to conserve their spatial structure both in the course of propagation in free space and during passage through lens-like media.

Substituting the Gaussian profile (1.158) in eqn (1.157), we obtain the function $M_\omega(\rho, z)$ relevant to this source (D'yakov 1993):

$$M_\omega(\rho, z) = \frac{\omega^2}{\omega^2 + \Omega^2} \exp\left[-\frac{\rho^2}{\rho_0^2}\frac{\omega^2}{\omega^2 + \Omega^2}\right], \tag{1.159}$$

$$\Omega = \frac{zc}{\rho_0^2 n(\omega)}. \tag{1.160}$$

According to eqn (1.159) the effective radius of the beam is

$$\rho(z) = \rho_0 \sqrt{\left(1 + \frac{z^2 c^2}{\rho_0^4 \omega^2 n^2(\omega)}\right)} \tag{1.161}$$

The dependence of the function M_ω on the frequency ω is shown in Fig. 1.21. The value $\omega = \Omega_m$ relates to the maximum of the function M_ω:

$$\Omega_m = \Omega\left[\left(\frac{\rho}{\rho_0}\right)^2 - 1\right]^{-1/2}, \quad M_\omega|_{\max} = \frac{\rho_0^2}{e\rho^2}. \tag{1.162}$$

If the distance from the source is small ($\Omega \ll \omega$), the dependence of the functian M_ω, eqn (1.159), on the frequency ω vanishes. Thus, for a laser generating radiation of wavelength $\lambda_0 = 1.06\,\mu$m, the aperture radius being $\rho_0 \approx 0.1$ cm, one can find from eqns (1.159)–(1.160) that the effects under discussion will become meaningful at distances $z \geqslant 5$ m in free space.

To evaluate the shift of the spectral maximum of the beam at some posi-

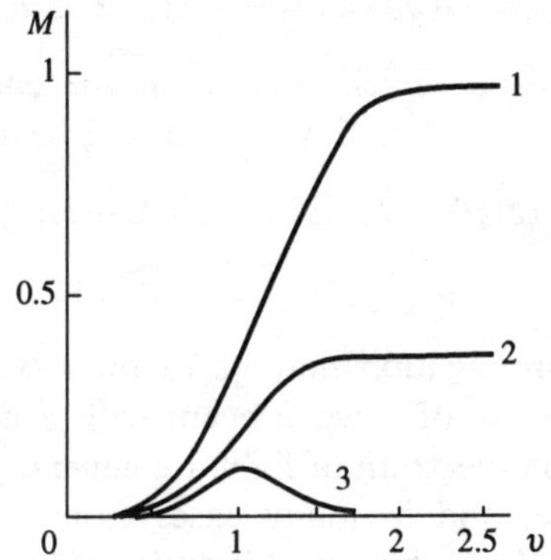

Fig. 1.21 Spectral transfer function $M\ (\rho, z)$ (eqn 1.159) for a nondispersive medium vs. normalized frequency $v = \omega\Omega_m^{-1}$, where Ω_m is given by eqn (1.162). Curves 1, 2, and 3 relate to ratio $\rho \cdot \rho_0^{-1} = 0$, 1 and 1.3 respectively.

tion z with respect to the spectral maximum of the source $G_0(\omega)\,|_{z=0}$, let us examine the Gaussian spectrum with width Ω_0:

$$G_0(\omega) = C \exp\left[-\frac{(\omega-\omega_0)^2}{\Omega_0^2}\right]. \tag{1.163}$$

Considering the far zone of the Gaussian source (1.159) ($\Omega^2 \gg \omega_0^2$) with the spectrum (1.163) in a nondispersive medium ($k = \omega c^{-1}$), one can find from eqn (1.156) for the beam axis ($\rho = 0$)

$$I_\omega(0, z) = \frac{C\omega^2}{\Omega^2} \exp\left[-\frac{(\omega-\omega_0)^2}{\Omega_0^2}\right]. \tag{1.164}$$

The spectral maximum of the field (1.164) relates to the frequency ω_{m},

$$\omega_{\mathrm{m}} = \frac{\omega_0}{2}\left\{1 + \sqrt{\left[1 + 4\left(\frac{\Omega_0}{\omega_0}\right)^2\right]}\right\}. \tag{1.165}$$

The analysis of formula (1.165) shows that the spectral maximum at the beam's axis is blue-shifted, while its largest value does not exceed the spectral width Ω_0. The shift of this maximum $\Delta\omega = \omega_{\mathrm{m}} - \omega_0$ at the beam's periphery can exceed the value Ω_0, and the sign of $\Delta\omega$ may be either positive or negative.

The accumulation of spectral changes due to diffraction depends on the structure of the wave beam. Thus Bessel beams are known to represent diffraction-free waves with a sharply defined central maximum (Durnin 1987). The amplitude profile of such a beam at the plane $z = 0$ is given by a zeroth-order Bessel function J_0:

$$E(\rho)|_{z=0} = J_0(a\rho)\exp(-\mathrm{i}\omega t), \quad 0 \leqslant a \leqslant \omega/c. \tag{1.166}$$

Unlike a Gaussian beam, which spreads in the course of propagation, eqn (1.161), the central spot radius of a Bessel beam remains fixed:

$$E(z,\rho) = J_0(a\rho)\exp[\mathrm{i}(\beta z - \omega t)], \tag{1.167}$$

$$\beta = \sqrt{[(\omega/c)^2 - a^2]}. \tag{1.168}$$

This distortion-free propagation may be of interest for reducing spectral distortions due to the use of a wave beam with a weakened divergence.

Measurements of the spectrum of light are generally made some distance away from its source, and in many cases (for example in astronomy) exceedingly far away. Thus this effect may be of interest for spectroscopy and optical metrology as well as for continuous tuning of frequency spectra of wideband wave sources through variation of source–target distances.

1.5.2 Smoothing of the diffraction pattern of a short pulse

Diffraction of a long train of monochromatic waves passing through an aperture is known to produce a spatial redistribution of this train, resulting in the formation of maxima and minima of intensity I on an illuminated surface. Thus the distributions of intensity on the image plane, parallel to the aperture plane, which are produced by diffraction of a monochromatic wave with the constant unit amplitude at slits (I_s) and circular apertures (I_c) are determined by the dimensionless functions (Born and Wolf 1964)

$$I_s = \left(\frac{\sin u_s}{u_s}\right)^2, \quad I_c = \left[\frac{2I_1(u_c)}{u_c}\right]^2. \tag{1.169}$$

Here

$$u_s = \frac{kax}{d}, \quad u_c = \frac{ka\rho}{d}, \tag{1.170}$$

where d is the distance between the aperture (A) and image (I) planes shown in Fig. 1.22, relevant to the diffraction at a long slit of width a, and x is the coordinate of the observation point $M(x)$ in the I plane; for diffraction at a circular opening, the variables a and ρ denote the radii of the opening and the observation point respectively. This picture is based on the stationary interference of fields irradiated by the different parts of the wave front passing through the aperture.

In contrast, the diffraction of a short pulse at a similar aperture is accompanied by nonstationary phenomena, violating the distributions (1.169). To discover some tendencies of such violations, let us reconsider for simplicity the intensity distribution (1.169) for a long slit (Fig. 1.22), keeping in mind

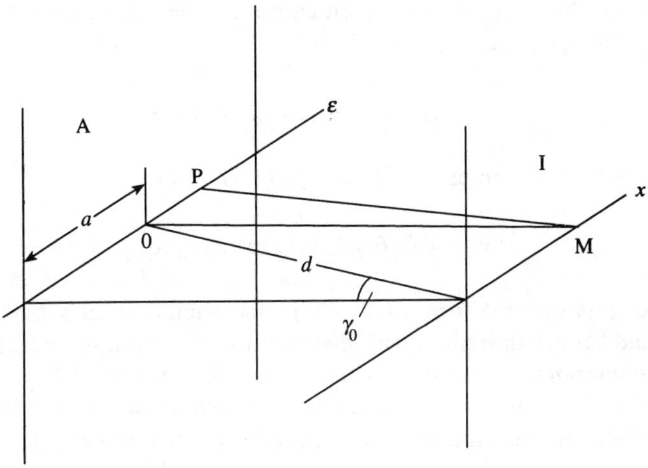

Fig. 1.22 Geometry of diffraction at a slit A and I are the aperture and image planes, M is the observation point, and $2a$ and d are the width of the slit and the distance between the A and I planes respectively; $\gamma_0 = a \cdot d^{-1}$.

that the delay time between the waves forming the interference pattern on a plane I is limited by the pulse duration T. Introducing the function $F(\gamma)$, $\gamma = \varepsilon d^{-1}$, describing the distribution of field amplitude $f(t)$ on a wave front across the slit, one can write the time-dependent amplitude $E(x, t)$ at some observation point $M(x)$ for small angles ($\gamma^2 \ll 1$) in the form

$$E(x, t) = \int f(t + \theta)F(\gamma)\exp([\mathrm{i}(k\gamma x - \omega t)]\,\mathrm{d}\gamma, \qquad (1.171)$$

$$\theta = \frac{\gamma x}{c}. \qquad (1.172)$$

Parameter θ determines the time delay between two waves arriving at point M from points O and P of the aperture. Considering the wave intensity at the point M, $I(x) = E(x, t)E^*(x, t)$, we obtain

$$I(x) = \int \mathrm{d}\gamma_1\,\mathrm{d}\gamma_2\,F(\gamma_1)F(\gamma_2)D(\theta_\mathrm{d})\exp[\mathrm{i}kx(\gamma_1 - \gamma_2)]. \qquad (1.173)$$

Here the function $D(\theta_\mathrm{d})$ is related to the delay parameter θ_d, eqn (1.172):

$$\theta_\mathrm{d} = \frac{x}{c}\gamma_\mathrm{d}, \quad \gamma_\mathrm{d} = \gamma_1 = \gamma_2 \qquad (1.174)$$

$$D(\theta_\mathrm{d}) = f(t)f^*(t - \theta_\mathrm{d}). \qquad (1.175)$$

Introducing the autocorrelation function of the amplitudes $F(\gamma_1)$ and $F(\gamma_2)$ at the aperture,

$$A(\gamma_d) = \int F(\gamma_1)F(\gamma_1 - \gamma_2)\, d\gamma_1, \tag{1.176}$$

one can rewrite the intensity $I(x)$, eqn (1.174), as

$$I(x) = \int D(\theta_d)A(\gamma_d)\exp(ikx\gamma_d)\, d\gamma_d. \tag{1.177}$$

To use this general formula (1.177), one should specify the functions $D(\theta_d)$ and $A(\gamma_d)$ describing the distribution of amplitude across the slit and the temporal envelope of the pulse. Remember that in the case discussed, the amplitude at the slit is supposed to be independent of the x-coordinate, so the function $F(\gamma)$ may be written as (Fig. 1.22)

$$F(\gamma) = \begin{cases} 1, & -\gamma_0 \leqslant \gamma \leqslant \gamma_0, \\ 0, & |\gamma| > \gamma_0, \end{cases} \qquad \gamma_0 = \frac{a}{d}. \tag{1.178}$$

Using the new dimensionless variable $y = \gamma_d \gamma_0^{-1} (-2 \leqslant y \leqslant 2)$, one can present the autocorrelation function $A(\gamma_d)$, eqn (1.177), in the form

$$A(y) = 1 - \frac{|y|}{2}. \tag{1.179}$$

As regards function $D(\theta_d)$, eqn (1.176), let us examine a truncated sine pulse with finite duration T and amplitude $f(t)$, normalized to unity:

$$f(t) = \begin{cases} 1, & -\dfrac{T}{2} \leqslant t \leqslant \dfrac{T}{2}, \\ 0, & |t| > T/2. \end{cases} \tag{1.180}$$

Since the amplitude $f(t)$ is constant, the function $D(\theta_d)$ may be written as

$$D(\theta_d) = 1 - \frac{|y\theta_0|}{2}, \quad \theta_0 = \frac{\gamma_0 x}{c}. \tag{1.181}$$

Substituting eqns (1.181) and (1.179) in eqn (1.177) and performing the integration with respect to variable y within the limits $(-2, 2)$, we obtain the intensity $I(x)$:

$$I(x) = I_s + \frac{1}{\omega_0 T} P(u_s), \tag{1.182}$$

$$P(u_s) = \frac{2\cos u_s}{u_s}\left(\cos u_s - \frac{\sin u_s}{u_s}\right). \tag{1.183}$$

Here the function I_s and the variable u_s are as defined by eqns (1.169)–(1.170).

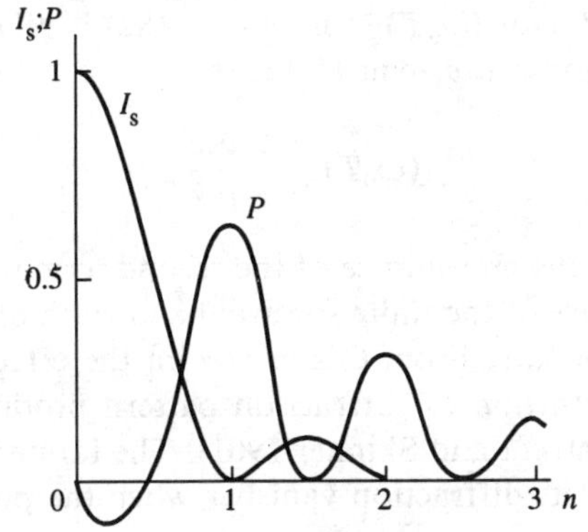

Fig. 1.23 Functions I_s (eqn 1.169) and P (eqn 1.183), describing the contributions of stationary (I_s) and time-dependent (P) diffraction effects to the diffraction pattern of a short square-shaped pulse (eqn 1.182); n is the normalized variable $n = u_s \cdot \pi^{-1}$; u_s is defined by eqn (1.170).

The distribution of intensity formed by diffraction of a short pulse is presented in eqn (1.182) as the sum of two terms: the first, coinciding with I_s in eqn (1.169), relates to the well-known diffraction pattern produced by a long train of monochromatic waves, whereas the second term describes the contribution of finite duration of a pulse to the intensity distribution. The functions I_s and $P(u_s)$ are shown in Fig. 1.23. The maxima of function $P(u_s)$ are located close to the minima of the stationary diffraction pattern, and vice versa; the main maximum of distribution I_s ($I = 1$) is not sensitive to the time-dependent effect. However, the other extremal points of the distribution I_s may be changed as a result of this effect, described by the function $P(u_s)$, eqn (1.183). The first maximum of $P(u_s)$ located near the point $u_s = \pi$, is $P_{max} \approx 2\pi^{-1}$; the contribution of this maximum to the diffraction pattern depends on the pulse duration T; for example, in the case of a single-cycle pulse $\omega_0 T = 2\pi$, the value of the second term in eqn (1.182) is ~0.1. Hence it appears that the first diffraction minimum at the point $u_s = \pi$ will increase from $I_s = 0$ to $I_s = 0.1$; this value is more than twice the first maximum of the unperturbed distribution $I_s(1.57\pi) = 0.045$. Thus the influence of the time-dependent phenomenon is to weaken the intensity modulation at the image plane and smooth the diffraction pattern.

Proceeding in a similar fashion, one can find the time-dependent corrections to the classical diffraction distribution in the case of a circular aperture also. The relevant formula may be obtained from eqn (1.182) through replacement of I_s by I_c, eqn (1.169), and numerical calculation of the function P in eqn (1.182) corresponding to the circular geometry. The values of the first maxima of these functions P, located close to the first minima of corrections for both slit and circular apertures, are of the same order of magnitude.

Noting that the factor $(\omega_0 T)^{-1}$ in eqn (1.182) is related to the spectral bandwidth of the pulse $\Delta\omega$, eqn (1.12), by

$$(\omega_0 T)^{-1} = \frac{\Delta\omega}{\omega k}, \tag{1.184}$$

one could interpret the appearance of the second term in eqn (1.182) as the result of the influence of the finite spectral bandwidth of such a short pulse on its diffraction picture. From this viewpoint the effect discussed resembles the similar distortion of diffraction pattern produced by a partially coherent source (Parrent and Skinner 1961). The influence of the duration of a short pulse on its diffraction vanishes when the pulse is long enough $(\omega_0 T \gg 1)$.

This wideband smoothing of the diffraction pattern may be of interest for precise optical processing of ultrashort signals containing only few oscillations of the wave field.

1.6 Resonant generation of transients by ultrashort waveforms

The discovery of reliable sources of picosecond single-cycle waveforms (Fattinger and Grischkowsky 1988) and femtosecond laser pulses (Spielmann *et al.* 1991) has opened up an interesting new avenue of research connected with the impulse excitement of electromagnetic eigenoscillations in resonant media. The re-emission of the excited eigenfrequencies can stimulate dramatic reshaping of the pulse reflected from these media, and formation of a long 'ringing' tail of the reflected pulse. The structure and spectrum of this tail, containing the above-mentioned eigenfrequencies, may provide valuable information about certain electromagnetic parameters of the reflecting medium.

Two physically meaningful mechanisms may stimulate the origin of such resonant phenomena:

1. Excitement of eigenfrequencies by some spectral components of the incidenting wideband pulse. The shorter the pulse and the wider its spectral bandwidth, the more likely will become such a resonant effect.
2. 'Shock' generation of exciton or molecular eigenoscillations in a transparent medium by the steepened fronts of the incident pulse. The reflected pulse is found to have transients associated with both its leading and trailing fronts.

'Shock' transient effects arise when the waveform has well-defined boundaries, such that the intensity becomes strictly zero at the waveform edges. Thus the widely spread models of Gaussian and hyperbolic-secant envelopes will not give rise to transients, owing to their smooth trend to zero at the edges; however, the finite spectral bandwidths of such pulses may be used for generation of resonant 'ringing' of reflecting media and scattering targets.

Unlike these transients arising from reflection of short pulses from resonant media, the propagation of wideband pulses within such media is known to give rise to the formation of another type of transient signals, so-called precursors (Brillouin 1960). If we assume that an incident pulse has a centre frequency close to the resonance of the material and a bandwidth wider than the absorption band, the resulting transmitted pulse will have an attenuated carrier and will therefore be dominated by the spectral components outside the absorption band. Since the part of the signal that is in the resonance region and is therefore characterized by a higher refractive index propagates more slowly than those parts outside this region, the outlying spectral components will reach an observer sooner than the main pulse does. The transient signal containing the spectral components above resonance is known as the Sommerfeld precursor; this signal passes through the medium first. Next comes the Brillouin precursor, which contains the frequencies below resonance. Finally the main body of the signal arrives. If the path of propagation is long enough, these three pulses will be separated distinctly (Oughstun 1991).

Some properties of transients formed by nonstationary reflection of a pulse from the plane boundary of a resonant crystal medium are discussed in Subsection 1.6.1., while Subsection 1.6.2 is devoted to the transient effects accompanying the scattering of short pulses on curvilinear targets when the resonant conditions depend upon the target's geometry.

1.6.1 Time-resolved reflectivity in terahertz optics

To consider the possibilities of observing transient reflection, one should use the model of a Lorentz resonant medium, described by the well-known formula for the complex dielectric permittivity $\varepsilon(\omega)$:

$$\varepsilon(\omega) = \varepsilon_\infty + \frac{\Omega_0^2(\varepsilon_s - \varepsilon_\infty)}{\Omega_0^2 - \omega^2 - \mathrm{i}\omega\Gamma + \hbar\Omega_0 k^2/M}. \tag{1.185}$$

Here Ω_0 and Γ are the eigenfrequency and damping factor of the elementary Lorentz oscillator, $\hbar$ is $h/2\pi$, where h is the Planck constant, $\varepsilon_\infty = \varepsilon(\infty)$, and $\varepsilon_s = \varepsilon(0)$. This model relates to a crystal, the spatial dispersion (nonlocality of the dielectric response) being taken into account; k is the wave-vector and M is the effective exciton polariton mass. By combining the expression (1.185) with the Fresnel formula for complex reflection coefficient R, we may use measurements of reflectivity for evaluation of parameters of the crystal. Thus the Fresnel coefficient R_p for a p-polarized plane wave may be written as

$$R_p(\omega, \gamma) = \frac{\varepsilon(\omega)\cos\gamma - \sqrt{[\varepsilon(\omega) - \sin^2\gamma]}}{\varepsilon(\omega)\cos\gamma + \sqrt{[\varepsilon(\omega) - \sin^2\gamma]}}, \tag{1.186}$$

where γ is the angle of incidence. Designating the spectral amplitude of an incident field as $E(\omega)$, we can calculate the reflected field $E_p(t)$ in the time domain by means of an inverse Fourier transform:

$$E_p(t) = \frac{1}{2\pi}\int_{-\infty}^{\infty} R_p(\omega)e^{-i\omega t}\,d\omega. \tag{1.187}$$

The nonzero spectral bandwidth of the incident field and the dispersion of the reflection coefficient $R_p(\omega)$ result in reshaping of the reflected waveform. This approach has been shown to be suitable for interpretation of different types of transients formed by interaction of picosecond pulses in the infrared range with GaAs and CdS crystals near excitonic resonances $\omega \approx \Omega_0$ (Agrawal *et al.* 1982).

Thus, in the simple case of a classical resonance ($M \to \infty$) and vanishing damping ($\Gamma = 0$), the observation of steady-state reflection spectra of a laser beam incident normal ($\gamma = 0$) to the surface of epitaxial GaAs samples revealed a characteristic excitonic dip in the vicinity of a frequency $\Omega_0 = 2 \times 10^{15}$ rad s^{-1} (Aaviksoo and Kuhl 1989). Irradiation of the same samples by hyperbolic-secant pulses with the same carrier frequency Ω_0 and a duration of 3.6 ps, and measurement of cross-correlation functions of the reflected pulses with the incident ones illustrated the formation of tails which extended beyond the initial pulse envelope. These phenomena may be important for the design of multilayer dielectric mirror coatings optimizing the distortion of reflected laser pulses.

To visualize the temporal reshaping of the reflected pulses, it may be expedient to use the reflection at an interface near the Brewster angle γ_B. This angle is determined by the condition that the reflection of a p-polarized wave from a lossless ($\Gamma = 0$) dielectric ($R_p = 0$) vanishes. Using the formula (1.186), one can find

$$\mathrm{tg}\,\gamma_B = \sqrt{\varepsilon(\omega)}. \tag{1.188}$$

Near this angle the phase and magnitude of the Fresnel coefficient R_p change rapidly with the angle. If a short laser pulse is incident close to Brewster's angle, the various frequency components of the pulse will be reflected differently, yielding reflected pulses that are reshaped. Thus, considering the model of a sech-shaped incident envelope,

$$E(t) = \frac{E_0\cos(\omega_0 t)}{\mathrm{ch}(t/T)}, \tag{1.189}$$

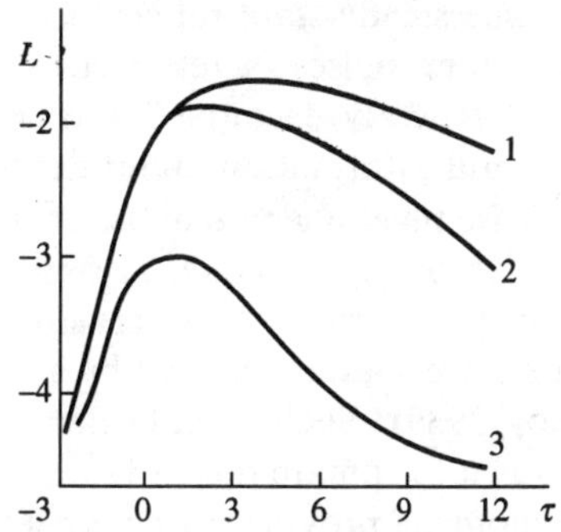

Fig. 1.24 Formation of a tail of a reflected P-polarized sech pulse at the Brewster angle of incidence near the excitonic resonance. $L = \lg(I \cdot I_0^{-1})$, $\tau = t \cdot T^{-1}$; I is the intensity of the tail at the moment t, I and T are the peak intensity and the temporal halfwidth the of incident pulse respectively. The values of the parameters in eqn (1.185) used in calculation of these graphs are $\varepsilon_\infty = 12.6$, $\varepsilon_S - \varepsilon_\infty = 4\pi \cdot 10^{-3}$, $\Omega = 55$ THz, $M = nm_e$, where m_e is the free electron mass. Curve 1 relates to the idealized model ($n = \infty$; $\Gamma = 0$), curves 2 and 3 illustrate the weakening of the time-resolved reflectivity through the influence of a finite damping factor ($\Gamma = 6$ GHz; $n = \infty$, curve 2) and through a finite value of exciton mass ($\Gamma = 0$, $n = 0.3$, curve 3).

substituting its Fourier transform

$$E(\omega) = \frac{E_0 T\pi}{2}\left\{\frac{1}{\mathrm{ch}[\pi T(\omega_0+\omega)/2]} + \frac{1}{\mathrm{ch}[\pi T(\omega_0-\omega)/2]}\right\} \qquad (1.190)$$

in eqn (1.187), and making use of the Fresnel formula (1.186), we can examine numerically the influence of damping factor Γ and nonzero exciton polariton mass M on the profile of the reflected pulse. The results of such calculations, depicted in Fig. 1.24, show that the duration of a ringing tail excited by the pulse (1.189) may be much longer than the characteristic pulse halfwidth T. The shorter the pulse halfwidth, the wider is the angular interval $|\Delta\gamma| = |\gamma - \gamma_B|$ associated with the nonzero reflection close to the Brewster angle. Thus the reflection of a 100 fs p-polarized IR pulse from a GaAs surface, characterized by the value $|\Delta\gamma| \leqslant 10^{-2}$ rad, results in strong reshaping of the reflected pulse, whereas for a 10 ps pulse the reshaping becomes negligible (Campbell and Fauchet 1988).

Side by side with these wideband phenomena, transient behaviour can be observed for pulses reflected from some semiconductors of the $A^{III}B^{V}$ group on a 1 ps time scale; such an effect is expected to arise when the exciting pulse possesses sharp temporal structures (leading or trailing edges) that are shorter than the characteristic reaction time of the system. A truncated square-shaped pulse is preferable for this 'shock' excitation near the excitonic resonance. It may be more convenient to look for the trailing-edge transient reflectivity, since steady-state reflectivity will then not interfere with measurements. For long pulses (a few picoseconds), transients from the two edges will be sufficiently decoupled and can be considered independently, presenting useful information about exciton parameters such as mass and decay rate. These phenomena may be of interest in the design of front-end receivers for servos and sensors. A similar effect of formation of 1 ps transients has been recorded by observation of terahertz fluorescence quantum beats in a dye solution excited by a femtosecond laser pulse (Mokhtari and Chesnoy 1989). Such rapidly damping transient beats of large molecules were shown to contain only a few oscillations with a picosecond period. These phenomena provide the physical basis for time-resolved fluorescence spectroscopy of molecular relaxation in some liquids. Moreover, the useful analogy between these optical effects and new trends in time-resolved diffraction of ultrashort electron and X-ray pulses should be mentioned.

It is also noteworthy that although the time-resolved reflection response is nothing but a Fourier transform of the reflection amplitude spectrum, it reveals the dynamics of the reflecting system in a straigtforward way. This effect provides a direct physical interpretation of some rapid processes in crystals, whereas the structure of the reflection spectrum in the frequency domain presents more indirect information. These shape-dependent reflections of ultrashort waveforms may be used in terahertz optics for controlled shaping of such waveforms, as well as for rapid remote sensing of certain solid surfaces (see Chapter 3).

1.6.2 Resonant reshaping of scattered signals

Electromagnetic fields of solitary ultrashort waveforms scattered on targets possess some characteristics distinguishing them from the fields of long trains of harmonic pulses scattered on the same targets. This distinction may be revealed clearly when the spatial scales of the interacting wave forms and scatterers become comparable. The time-domain response of a scatterer often contains useful information about the nature of scattering, in an easily interpretable form. This response can be decomposed into two components: the 'early time' echo, which is the result of the incident signal being scattered by the target skin (which responds at about the speed of light), and the 'late time' or natural response, resulting from the target ringing after the impulse has passed. Obviously a complex target will have many different resonances; some of them can be excited by the frequencies from the wide spectrum of the ultrashort waveform. The natural response depends solely on the target and is independent of the incident angle and polarization. These 'early time' and 'late time' contributions are often separated distinctly in time; hence they can be identified conveniently. Unlike the resonant phenomena dictated by excitonic eigenfrequencies of the medium, discussed above (Subsection 1.6.1), we consider here the scattering on targets produced from non-resonant materials, where the resonant scattering is related to the size and geometry of the target.

To compare the scattered fields of broadband pulses containing only a few cycles, let us proceed to the scattering of a plane wave pulse E_{in} on spheres with the same radius a but different reflectivities. Considering for simplicity the backscattering, one can use the Mie approach (Mie 1908), representing the spectral amplitude of the scattered field as a sum of contributions relevant to the multipoles induced by the incident field on the target

$$\left.\begin{aligned} E_{\text{bs}}(\omega) &= \frac{E_0}{kr} F(0) \exp[-\mathrm{i}(kz+\omega t)], \\ F(0) &= -\mathrm{i}\sum_{n=1}^{\infty} (-1)^n (n+1/2) M_n. \end{aligned}\right\} \tag{1.191}$$

Factors M_n are determined by the Mie coefficients, depending on the parameter $ka \leqslant 1$ and the structure of the scatter; thus the factor M_1 relates to dipole scattering (Rayleigh approximation) and so on.

Summarizing the spectral amplitudes, eqn (1.191), we obtain the backscattered field for a wideband pulse in the form of Fourier integral:

$$E_{\text{bs}} = \frac{E_0}{2\pi r} \int \frac{\mathrm{d}x}{x} G(k) F(0) \exp(-\mathrm{i}x\tau). \tag{1.192}$$

Here $G(k)$ is the spectrum of the incident pulse, $x = ka$, and τ is the dimensionless time

$$\tau = \frac{ct - r}{a}. \tag{1.193}$$

Selecting for simplicity a truncated sinusoidal pulse of finite extent with carrier frequency ω_0,

$$E_{\text{in}}(t)\,|_{z=0} = \begin{cases} \sin(\omega_0 t), & 0 < t \leqslant t_0, \\ 0, & t > t_0, \end{cases} \tag{1.194}$$

one can find the spectrum $G(k)$ in eqn (1.192) in a form similar to eqn (1.9):

$$G(k) = \frac{a}{2}\left[\frac{1 - \mathrm{e}^{\mathrm{i}(x+x_0)T}}{x + x_0} - \frac{1 - \mathrm{e}^{-\mathrm{i}(x-x_0)T}}{x - x_0}\right], \tag{1.195}$$

where $x_0 = a\omega_0 c^{-1}$, $T = ct_0 a^{-1}$. Substituting the spectrum $G(k)$, eqn (1.195), in eqn (1.193), we obtain the normalized backscattered field $e_s = E_{\text{bs}} E_0^{-1}$.

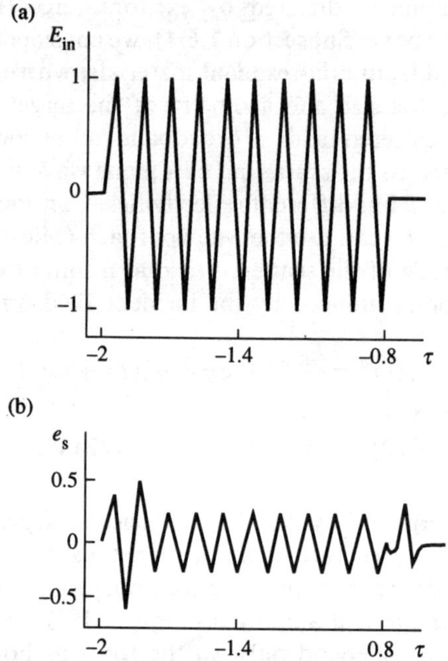

Fig. 1.25 (a) Amplitude of an incident ten-cycle pulse E and (b) normalized amplitude of the backscattered pulse e_s, vs. the dimensionless time variable τ (eqn 1.193). The parameters of the incident pulse (eqn 1.195) are $x_0 = 52.5$, $T = 1.2$; the backscattered pulse proves to be longer than the incident one.

Let us consider first the transient echo from a perfectly conducting sphere (Fig. 1.25). The reflected pulse starts out at $\tau = -2$, eqn (1.193), owing to its reflection from the sphere's surface ($r = a$; $ct = -a$). Using for the calculations the model of a ten-cycle pulse with a free-space wavelength $\lambda_0 = 118\,\mu$m (methanol laser), scattered on a sphere of radius $a = 0.1$ cm, one can find the dimensionless parameters in eqn (1.195): $x_0 = 52.5$; $T = 1.2$. Comparing the incident and backscattered pulses, we can see that the backscattered one is followed by a short tail, the duration $\Delta\tau$ of which is shorter than that of the incident pulse T, eqn (1.195). This tail constitutes the damping 'ringing' of the sphere due to excitation of some of its natural electromagnetic resonances.

This tendency to expansion of a scattered pulse can be visualized by consideration of an opposite case: a perfectly transparent sphere with radius a and refractive index $n > 1$ (the dispersion may be neglected in this example). Considering a ray with wavenumber k passing outside the sphere and refracted into the sphere, one should examine the trajectories of refracted rays localized within the subsurface layer (Fig. 1.26)

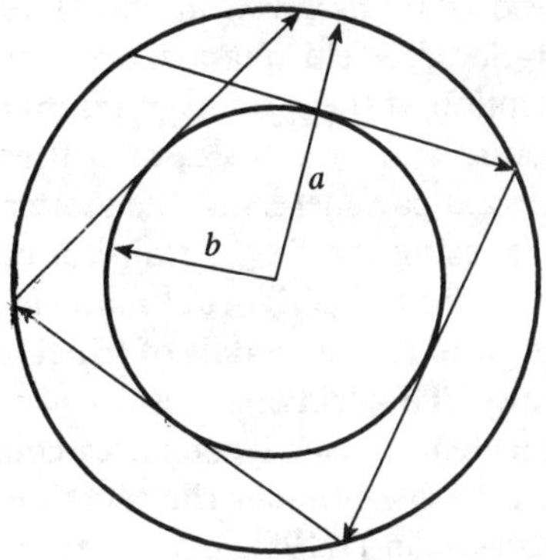

Fig. 1.26 Geometry of tunnelling of a wave incident on a transparent dielectric sphere with radius a and refractive index n; the tunnelling of a wave trapped in the spherical layer between the radius $\tau = a$ and $\tau = b$ (eqn 1.196) is associated with multiple internal reflection.

$$a > r > b = \frac{1}{k \cdot n}, \qquad (1.196)$$

where l are the discrete values ($l = m + 1/2$; $m = 0,1,2,\ldots$) of the angular momenta of EM eigenoscillations of the sphere. These trajectories of light rays around the scatterer's interior may be interpreted as resonant quasi-bound states of the wave field. The multiple reflections of trapped rays travelling inside the layer (eqn 1.196) are characterized by almost total reflection of their power fluxes; the leakage of energy to the outside as a result of each internal reflection is small.

Because of the leakage, these resonant states are not stationary, but their lifetimes, corresponding to many internal reflections, may be long enough. In view of the high angular momenta carried by the resonant states, and their confinement to a relatively narrow neighbourhood of the surface, these states resemble the orbiting phenomena in quantum mechanics and whispering-gallery modes in acoustics; moreover, these states can be considered as slowly-damping creeping waves on the sphere's surface. Introducing the radial wavenumbers K_{ef} of the trapped rays,

$$K_{ef}(r) = \sqrt{[k^2n^2 - (l/r)^2]}, \qquad (1.197)$$

one can use a WKB-type approximation to evaluate the number p of excited states:

$$\int_z^b K_{ef}(r)\,\mathrm{d}r = \left(p + \frac{1}{2}\right)\pi, \quad p = 0,1,2,\ldots \qquad (1.198)$$

where the values a and b are as given in eqn (1.196). Formula (1.198) is analogous to a Bohr–Sommerfeld quantization condition.

An analytical description of the systems of resonances and their lifetimes, determining the structure and time scales of scattered pulses by means of the geometry and physical parameters of the scatterer, has been developed by Nussenzveig (1992), using the Regge trajectories and complex poles of the S-matrix. As distinct from this rigorous mathematical theory, we would emphasize here the physical fundamentals of 'ringing' of scatterers after the passage of a short pulse. This 'ringing' results in the formation of a pronounced tail if the incident carrier frequencies coincide with some eigenfrequencies of the target. The broader the spectrum of the incident pulse, the more resonant states, eqn (1.198), can be excited, dictating the length and the spectra of the 'ringing' tail of the pulse. On the other hand, the consideration of a transparent sphere as an optical cavity draws attention to the possibilities of lasing from, for example, a solid ND:YAG sphere.

Considering the pulse excitation of a transparent microcavity containing some resonant medium, one should mention the influence of the geometry of this cavity on the shift of eigenfrequencies of the resonant medium. This effect may be illustrated by the model of a solid plasma specimen, e.g. semiconductor microdroplets. Thus the reflection of short waveforms from the interface of the plasma half-space stimulates the effect of 'plasma ringing' (Luebbers *et al.* 1991) in the vicinity of the Langmuir frequency ω_p:

$$\omega_p = \sqrt{\left(\frac{4\pi e^2 N}{m}\right)}, \tag{1.199}$$

where e, m and N are the charge, mass and density of the electrons in the plasma. On the other hand, the frequency Ω_e for the semiconductor plasma specimen may be changed as a result of the depolarization effect induced by the alternating electric field. Thus, the value Ω_e for the electrons with an effective mass m^* in an InSb spheroid depends on the depolarization factor p and axial ratio q (Dresselhaus *et al.* 1955):

$$\Omega_e = \sqrt{\left(\frac{4\pi e^2 N}{m^*}\frac{p}{1+pq}\right)}.$$

In the case of an unbounded plasma medlum ($p = 1$, $q = 0$), this value tends to the value ω_p, eqn (1.199).

Keeping in mind the tendencies discussed above for pulse scattering on both perfectly conducting and perfectly transparent spheres, one could consider an important intermediate case of a heterogeneous scatterer. Consideration of such a scatterer, modelled by a perfectly conducting sphere coated with a transparent dielectric layer, has revealed several separated damping bursts of eigenfrequencies in its tail (Howell and Uberall 1990). Similar natural resonances can be excited by transients at scatterers of geometries, for example in scattering of single-cycle waveforms by a thin finite wire (Guo and Uberall 1994). This method, developed initially in the radio range, is now opening up new horizons in the optics of ultrashort solitary waveforms.

1.7 Shape-dependent absorption of transients

The recent successes in generation and detection of freely propagating electromagnetic waveforms in the picosecond and femtosecond regions have given rise to the investigation of impulse interactions of such fields with dielectrics in the time domain. The use of photoconductive switching and sampling (Downey and Tell 1984) as well as pulsed optoelectronic antennas (Pastol *et al.* 1989) has demonstrated the feasibility of a form of transient spectroscopy for measurements of absorption coefficients and refractive indices of some dielectrics in the fields of single-cycle picosecond pulses. The duration of these pulses becomes comparable with the characteristic times of rotational and some vibrational transitions of molecules, $t_0 = \omega_{mn}^{-1}$; where ω_{mn} are the frequencies of such transitions in the far infrared or submillimetre regions. This circumstance opens up new possibilities for ultrafast spectroscopy of molecular states. On the other hand, the time scales τ of pulses generated by femtosecond laser systems may be much shorter than ω_{mn}^{-1}. These conditions,

$$\tau\omega_{mn} \ll 1, \tag{1.200}$$

change in principle the usual mechanism of interaction of light with multilevel quantum systems. Compared with traditional resonant excitation of molecular levels by long trains of monochromatic waves, this transient excitation has been shown to result in the development of unusual nonresonant, nonstationary phenomena in such systems.

To visualize these ultralast phenomena, the time-domain description is used below for analysis of both shape-dependent absorption of waveforms in two-level molecular systems and formation of nonequilibrium states in three-level ones.

1.7.1 Time-resolved absorption spectroscopy of non-sinusoidal waveforms

Let us examine the interaction of electric waveforms $E(t) = E_0 U(t)$ with a two-level molecule which may be in either of two states of definite energy ε_1 or ε_2 such that $\varepsilon_1 - \varepsilon_2 = \hbar\omega_0$, where $\hbar = h/2\pi$, h being the Planck constant. Several decades ago this model was the subject of speculation connected with the physics of the ammonia maser, based on harmonic pumping of NH_3 molecules $[U(t)=\exp(-\mathrm{i}\omega t)]$. To reconsider this problem for a nonsinusoidal waveform $U(t)$, one should start from the system of equations of quantum mechanics (Feynman 1972), describing the behaviour of two-level system in an external time-dependent electric field $E(t)$:

$$\left.\begin{aligned} \mathrm{i}\hbar\frac{\partial\gamma_1}{\partial t} &= \mu E_0 U(t)\,\mathrm{e}^{\mathrm{i}\omega_0 t}\gamma_2 \\ \mathrm{i}\hbar\frac{\partial\gamma_2}{\partial t} &= \mu E_0 U(t)\,\mathrm{e}^{-\mathrm{i}\omega_0 t}\gamma_1 \end{aligned}\right\} \tag{1.201}$$

where μ is the electric dipole moment of the transition and γ_1 and γ_2 are the amplitudes of transitions between the states $1 \to 2$ and $2 \to 1$ respectively; the relevant probabilities are known to be $|\gamma_1|^2$ and $|\gamma_2|^2$. Supposing that at $t=0$ our molecular system is in the fundamental state I ($\gamma_1=1$, $\gamma_2=0$), we may discuss an important case of resonant pumping, $U(t)=\cos(\omega_0 t)$, by a long train of monochromatic waves. Substituting this function $U(t)$ in eqn (1.201) and omitting the terms in $2\omega_0$, which oscillate very rapidly about zero, me obtain the well-known result for time-dependent probabilities $|\gamma_1|^2$ and $|\gamma_2|^2$, the electric field E_0 being weak:

$$\Omega=\frac{\mu E_0}{\hbar}\ll\omega_0, \tag{1.202}$$

$$|\gamma_1|^2=\cos^2(\Omega t),\quad |\gamma_2|^2=\sin^2(\Omega t). \tag{1.203}$$

Thus the probability of being in the fundamental state $|\gamma_1|^2$ changes from 1 to 0 and back again, while the probability of an inverse transition oscillates as $|\gamma_2|^2=1-|\gamma_1|^2$. Generalizing this analysis, one can consider in the framework of eqn (1.201) an anharmonic pumping $U(t)$ also. Remembering the description of arbitrary waveforms by means of Hermitian functions (Subsection 1.2.2), let us discuss the absorption of odd and even wave forms represented by the Hermitian envelope Φ_1, eqn (1.30), (Fig. 1.9) and the 'Mexican hat', eqn (1.37) (Fig. 3.2), with time scale T. Suppose that the molecule is in the fundamental state before the incidence of the waveform at the moment $t\ll -T_0(t\to 0)$:

$$\gamma_1|_{t\to-\infty}=1,\quad \gamma_2|_{t\to\infty}=0. \tag{1.204}$$

The electric field E_0 being weak, eqn (1.202), the probability $|\gamma_1|^2$ of the molecule remaining in the fundamental state during the time $t\gg 2T_0$ is expected to be nearly equal to 1, whereas $|\gamma_2|^2$ will be small compared with unity. Substituting the Hermitian waveforms in eqn (1.201) and performing the integration over t between the limits $(-\infty<t<\infty)$, we get the probability

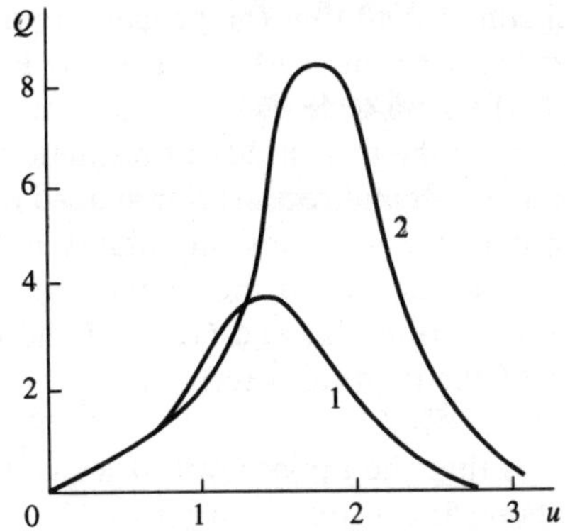

Fig. 1.27 Shape-dependent absorption of transients with time scale T in a two-level system with frequency of transition ω_0. The function Q (eqn 1.206), determining the probability of transition, is plotted vs. parameter $u = \omega_0 T$; curves 1 and 2 correspond to the Hermitian waveform $\Phi_1(t)$ (eqn 1.31) and the 'Mexican-hat' waveform F (eqn 1.38) respectively.

$$|\gamma_2|^2 = \left(\frac{\mu E_0}{\hbar\omega_0}\right)^2 Q(\omega_0 T_0), \tag{1.205}$$

$$Q(\omega_0 T_0) = (\omega_0 T_0)^2 |U_\omega|^2. \tag{1.206}$$

Here U_ω is the Fourier transform of the waveform profile $U(t)$. The functions $Q_{1,2}(\omega_0 T_0)$ corresponding to the aforesaid profiles $U_1 = \Phi_1$ (eqn 1.30) and F (eqn 1.37) may be calculated by means of the formulae (1.34)–(1.35):

$$\left.\begin{aligned} Q_1 &= 4\sqrt{\pi}\,(\omega_0 T_0)^4 \exp[-(\omega_0 T_0)^2], \\ Q_2 &= 2\pi(\omega_0 T_0)^6 \exp[-(\omega_0 T_0)^2]. \end{aligned}\right\} \tag{1.207}$$

The dimensionless parameter $(\mu E_0)^2(\hbar\omega_0)^{-2}$ in eqn (1.205) is smalls; thus, for typical molecular far-IR spectra ($\mu = 1$ D, $\omega_0 = 10^{14}$ rad s^{-1}) we obtain even at a field strength $E_0 = 3 \times 10^3$ V m^{-1} a value of this parameter of $\sim 10^{-1}$.

Keeping the transition frequency ω_0 fixed, one can discover the dependence of functions Q on the shape of the waveform and its time scale T. The curves Q, eqn (1.205), depicted for some Hermitian waveforms in Fig. 1.27, illustrate the dependence of the probability of transition $|\gamma_2|^2$ on the time scale T_0; when T_0 is fixed, this probability is shown to depend on the waveform's envelope. Figure 1.27 demonstrates some essential distinctions between the standard theory of light absorption by molecules and the developing concept of absorption of nonsinusoidal transients by two-level quantum systems:

1. It follows from eqn (1.205) that the probability of transition is proportional, as usual, to the peak intensity of the light. However, as regards dependence of $|\gamma_2|^2$ on the time scale T_0, the situation is not so trivial. Let us compare eqn (1.205) with the probability of resonant transition induced, unlike eqn (1.203), by a broadband rectangular-shaped pulse of finite duration $2T_0$ and carrier frequency ω_0. This probability is known to increase monotonically with the extension of pulse duration. In contrast, the dependence of $|\gamma_2|^2$ on the time T_0, eqn (1.206) is nonmonotonic; hence by choosing the shape of the exciting waveform $U(t)$ we can optimize the time scale T_0.

2. Figure 1.27 shows that the upper level of a two-level molecule may be excited by transients during a time as short as $(2-3)\omega_0^{-1}$. To provide this excitation by resonant quasi-monochromatic pulses with carrier frequency ω_0 and small spectral bandwidth $\Delta\omega(|\Delta\omega|\omega_0^{-1} \ll 1)$, long pulses with time scales $T \gg \omega_0^{-1}$ are required. Using the width of the absorption line ν associated with this transition, one can limit the values of spectral bandwidth of the light $\Delta\omega$ by the condition $|\Delta\omega| \approx \nu$. Therefore the duration of such resonant pulse must be $T \approx \nu^{-1}$. The natural width of vibrational transitions between adjacent levels of molecules in the infrared range may be as small as $\nu \approx$ 200–500 Hz (Demtroder 1982); hence the relevant pulse duration proves to be $T \approx 10^{-3}$ s, which is much longer than the transient's time scale $T_0 \approx \omega_0^{-1}$ ($\omega_0 \approx 10^{14}$ rad s^{-1}).

3. The dependence of the absorption of electric transients on their shapes demonstrates the possibility, together with the shape-dependent reflection of such transients from a dispersive interface (Chapter 3), of passive filtration of waveforms with respect to their envelopes.

To evaluate the effectiveness of the attenuation of the flux of transients as a result of the effect discussed, one can determine an energy $\mathcal{E}_1$ absorbed in unit volume of a lossy medium as $\mathcal{E}_1 = \hbar\omega_0|\gamma_2|^2 N$, where N is the density of the absorbing molecules. The density of electromagnetic energy $\mathcal{E}_2$ of the waveform $E(t) = E_0 U(t)$ may be represented via an averaged field amplitude E_a:

$$\mathcal{E}_2 = \frac{E_a^2}{4\pi}, \quad E_a^2 = E_0^2 \mathcal{K}, \quad \mathcal{K} = \frac{1}{2T_0}\int_{-\infty}^{\infty} U^2(t)\,dt, \tag{1.208}$$

where T_0 is the effective width of the waveform. Thus the ratio $\gamma = \mathcal{E}_1 \mathcal{E}_2^{-1}$ may be written as

$$\gamma = \frac{4\pi\mu^2 NQ}{\hbar\omega_0 \mathcal{K}}. \tag{1.209}$$

Here the function Q is given by eqn (1.208). Considering a molecular density $N \approx 10^{19}$ cm^{-3} and $\omega_0 = 10^{14}$ rad s^{-1}, and using the dimensionless factor K

for the Hermitian waveform $U = \Phi_1(t)$ (eqn 1.30) ($\mathcal{K} = 4\pi^{-1/1}$), one can find the ratio $\gamma = 10^{-3}$; the value of γ decreases in proportion to the density of the absorbing molecules. Therefore receivers of waveforms must be sensitive enough to record such small variations in energy fluxes.

Hence an important property of this transient absorption is associated with the rigid excitation of molecular transitions, although the probabilities of such transitions prove to be small compared with the slower but powerful excitations produced by resonant sinusoidal pulses. These phenomena may be of interest for the development of time-resolved absorption spectroscopy of some ultrafast kinetic processes, such as photochemical reactions and molecular relaxation.

1.7.2 Interaction of modulated pulses with multilevel quantum systems

Generalization of the results obtained for the two-level problem in Subsection 1.7.1 to transitions in multilevel systems is trivial, as long as these transitions are not forbidden by selection rules. However, such rapid phenomena in quantum systems may become much more intricate when certain dipole transitions are forbidden. The transient excitation of such transitions via the virtual level may provide unusual nonresonant formation of nonequilibrium states in such systems.

Let us suppose that the dipole transition between the levels n and m is forbidden by selection rules ($\mu_{nm} = 0$); however, the two-quantum transitions via some kth virtual level prove to be possible. We examine the excitation produced by a short modulated pulse with a carrier frequency ω_0 and a time scale of the envelope T_0. Unlike the treatment in Subsection 1.7.1, focused on the case $T_0\omega_{nm} \approx 1$, the case $T_0\,\omega_{nm} \ll 1$ will be discussed below. Since the amplitude of the dipole transition a_m is supposed to be zero in the first-order perturbation theory, one can use the second-order approximation. This model describes three groups of processes:

(1) absorption of two photons $\hbar\omega_0$;

(2) radiation of two photons $\hbar\omega_0$;

(3) consecutive absorption and radiation (or vice versa) of photons $\hbar\omega_0$.

The frequency ω_0 is not resonant with the frequency of transition between an initial nth and any virtual kth level. Omitting rapidly oscillating terms, containing the factors $\exp[\mathrm{i}(\omega_{nm} + 2\theta_0)t]$ relevant to the above-mentioned effects (1) and (2), one can represent the coherent part of the transition amplitude, describing process (3), in the form

$$a_m^{(2)}(\infty) = -\frac{\mathrm{i}E_0^2}{4\hbar^2}\int_{-\infty}^{\infty} q_{mn}\,U^2(t)\exp[\mathrm{i}\omega_{mn}t]\,\mathrm{d}t, \qquad (1.210)$$

$$q_{mn} = \sum_k \mu_{nk}\mu_{km} \left(\frac{1}{\omega_{km} - \omega_0} + \frac{1}{\omega_{kn} + \omega_0} \right) . \qquad (1.211)$$

Thus the forbidden transition may be excited by means of an ultrashort modulated pulse, the transition amplitude being proportional (not as in the case $T\omega_0 \approx 1$) to the square of the field's amplitude. The condition $T\omega_0 \ll 1$ permits the expressions containing this amplitude to be simplified by replacing the exponental factor, e.g. $\exp(\mathrm{i}\omega_{mn})$ in eqn (1.210), by unity.

It is worth while now to use these results to consider the pulse excitation of a multilevel system containing both permitted and forbidden transitions. To visualize some physically meaningful results, we may restrict ourselves to a three-level system with energy of levels $\mathcal{E}_1$, $\mathcal{E}_2$, and $\mathcal{E}_3$; the dipole transitions $1 \rightleftharpoons 2$ and $2 \rightleftharpoons 3$ are permitted by the selection rules, but the transition $1 \rightleftharpoons 3$ is forbidden, for instance as a result of the same parity of the states 1 and 3. The condition $T_0\omega_{ij} \ll 1$ is assumed to be fulfilled for states 1, 2, 3, while the other states are located far from these. The equations for the amplitudes of the probabilities of electrons existing at the levels 1, 2 or 3 are (Belenov *et al.* 1993)

$$\mathrm{i}\hbar \frac{\partial a_1}{\partial t} = -\mu_{12} E(t) a_2 - q_{13} E^2(t) a_3, \qquad (1.212)$$

$$\mathrm{i}\hbar \frac{\partial a_2}{\partial t} = -\mu_{21} E(t) a_1 - \mu_{23} E^2(t) a_3, \qquad (1.213)$$

$$\mathrm{i}\hbar \frac{\partial a_3}{\partial t} = -\mu_{32} E(t) a_2 - q_{31} E^2(t) a_1. \qquad (1.214)$$

Here μ_{ij} are the dipole moments of the $i \rightleftharpoons j$ transitions, and $q_{13} = q_{31}$ is the matrix element of second order (eqn 1.211) describing the transition $1 \rightleftharpoons 3$ via the virtual levels. Let us suppose that before switching of the exciting pulse, the system is in the state 1:

$$a_1(-\infty) = 1, \quad a_2(-\infty) = a_3(-\infty) = 0. \qquad (1.215)$$

The transitions $1 \rightleftharpoons 2$ and $3 \rightleftharpoons 2$ are stimulated by a rapidly oscillating field $E(t)$ (eqn 1.213) with average value $E(t)$ close to zero, unlike the transitions $1 \rightleftharpoons 3$, characterized by an average nonzero power-dependent effect $E^2(t)$. Thus the electrons are expected to be redistributed at levels 1 and 3. To find this redistribution, we can first integrate eqn (1.213), keeping in mind that the amplitudes a_1 and a_3 vary slowly compared with the change in the field $E(t)$:

$$a_2(t) = \frac{\mathrm{i}E_0}{\hbar} [\mu_{21} a_1(t) + \mu_{23} a_3(t)] \int_{-\infty}^{t} U(t)\, \mathrm{d}t. \qquad (1.216)$$

Substitution of eqn (1.216) in eqns (1.212) and (1.214) yields the equations governing the slowly varying amplitudes $a_1(t)$ and $a_3(t)$:

$$\left.\begin{aligned} i\hbar\frac{\partial a_1}{\partial t} &= -q_{13}E_0^2U^2(t)a_3, \\ i\hbar\frac{\partial a_3}{\partial t} &= -q_{31}E_0^2U^2(t)a_1. \end{aligned}\right\} \tag{1.217}$$

This system is independent of any parameters characterizing the transitions $1 \rightleftharpoons 2$ and $2 \rightleftharpoons 3$. The solution of the system (1.217), obeying the initial conditions (1.215), may be written as

$$\left.\begin{aligned} a_1(t) = \cos\Psi(t), \quad a_3(t) = i\sin\Psi(t), \\ \Psi(t) = \frac{q_{13}E_0^2}{2}\int_{-\infty}^{t} U^2(t)\,dt. \end{aligned}\right\} \tag{1.218}$$

The redistribution of electrons between levels 1 and 3 after the passage of the exciting pulse is determined by the probabilities n_1 and n_3 describing the nonequilibrium states:

$$\left.\begin{aligned} n_1 = \cos^2[\Psi(\infty)], \quad n_3 = \sin^2[\Psi(\infty)], \\ \Psi(\infty) = \frac{q_{13}E_0^2}{2}\int_{-\infty}^{\infty} U^2(t)\,dt. \end{aligned}\right\} \tag{1.219}$$

If the energy of the pulse is chosen so that

$$\Psi(\infty) = \pi(n + 1/2), \quad n = 0, 1, 2, \ldots, \tag{1.220}$$

the system will be converted exactly to the upper state $[n_1(\infty) = 0, n_2(\infty) = 0, n_3(\infty) = 1]$. The energy may be extracted from this nonequilibrium system by the use of two pulses relevant to resonant transitions $3 \to 2$ and $2 \to 1$.

Thus the interaction of a short modulated pulse with a quantum system is characterized by a peculiar type of selectivity that is not due to utilization of the resonant properties of this system. This rapid interaction does not excite the permitted transition, whereas the forbidden transitions prove to be excited effectively. This selectivity mechanism is found to be efficient in the formation of nonequilibrium states in multilevel quantum systems. This effect may be of interest for fast pumping of vibrational states of molecules and their dissociation by femtosecond optical pulses.

1.8 Time-domain and frequency-domain optics of short waveforms

The peculiarities of pulse phenomena in the optics of dispersive media may sometimes be considered as the wideband radiation effects described in the frequency domain by the Fourier transform. However, some such effects are also typical of the continuous fluxes of radiation with finite spectral bandwidth, for example the diffractive variations of the spectrum of the source analysed above in Subsection 1.5.1. On the other hand, short pulses may have vastly different effects from smooth wideband waveforms in both reflection and propagation processes. To reveal these effects, the time-domain treatment may be preferable.

The dynamics of a localized electromagnetic field in the time domain is traditionally examined either by the inverse Fourier transform of a result first obtained in the frequency domain, or by numerical solution of the Maxwell equations. Examples of exact inverse Fourier transforms in pulse optics of dispersive media can be found only rarely, and only for special choices of initial pulse shape and dispersion relation. Thus the dispersive spreading of a Gaussian modulated pulse travelling in a lossless plasma close to the cutoff frequency Ω_e may be treated in such a way. Let us assume that the initial shape of such a pulse at some moment $t = 0$ *is*

$$u(z)\big|_{t=0} = \exp\left(-\frac{z^2}{2L^2}\right)\cos(k_0 z). \tag{1.221}$$

The Fourier amplitude of this pulse, calculated by analogy with eqn (1.16), may be written as

$$u(k) = \frac{L}{2}\left\{\exp\left[-\frac{L^2(k-k_0)^2}{2}\right] + \exp\left[-\frac{L^2(k+k_0)^2}{2}\right]\right\}. \tag{1.222}$$

The plasma dispersive equation in the vicinity of the cutoff frequency Ω_e is known to be

$$k^2 = \frac{2\Omega_e(\omega - \Omega_e)}{c^2}. \tag{1.223}$$

Performing an inverse Fourier transform of eqn (1.222) and making use of eqn (1.223), we obtain an explicit expression describing the Gaussian pulse at an arbitrary time $t \geqslant 0$ and arbitrary point $z \geqslant 0$ within the plasma (Jackson 1975):

$$F = \mathrm{Re}\left\{\left(1 + \frac{\mathrm{i}t}{\mathrm{k_0 v_g}T^2}\right)^{-1}\exp\left[\mathrm{i}(k_0 z - \omega t) - \frac{(z - v_g t)^2}{2L^2}\right]\right\}. \tag{1.224}$$

Here the wave-vector $k_0 = k(\omega_0)$ relevant to the carrier frequency ω_0 is defined by eqn (1.223), and v_g is the group velocity of the haxmonic ω_0:

$$v_g = \frac{c^2 k_0(\omega)}{\omega_0}. \tag{1.225}$$

The salient features of propagation regime (1.224) are:

1. The peak amplitude of the pulse travels with the group velocity v_g, which tends to zero when the carrier frequency ω_0 tends to the cutoff frequency Ω_e.
2. The envelope remains Gaussian in shape.
3. The width of the envelope increases with time: $T = T_0 \mathcal{K}$,

$$\mathcal{K} = \sqrt{\left[1 + \left(\frac{t}{T_0} \frac{1}{k v_g T_0}\right)^2\right]}, \tag{1.226}$$

   whereas the pulse peak becomes damped with time: $F_{max} \propto \mathcal{K}^{-1}$.

The initial shape of the travelling waveform here is fixed not at the plasma boundary but at an arbitrary point within the plasma. This approach, justified for a Gaussian pulse because its peculiar properties (1)–(3), is ineffective for other pulses which distort in the course of propagation. The initial shape of these pulses must be specified at an arbitrary moment t at the boundary of the dispersive medium, $z = 0$. This requirement will be shown to present a series of problems in the Fourier approach to the dynamics of short waveforms in dispersive media.

Some of these problems may be avoided by direct integration of the Maxwell equations in the time domain. This method is based on replacement of the differential equations of electrodynamics by finite-difference equations solved by computer. Continuous space–time distributions of electromagnetic fields are approximated in the framework of such an approach by their values at some discrete space points and time points. This method, developed in computational electromagnetics beyond the scope of the Fourier approach, provides accurate numerical models for transient effects in real dispersive, lossy and layered structures used in quantum electronics. An example of the application of this method to a one-dimensional problem is discussed in Subsection 1.8.1; Subsection 1.8.2 indicates the principal problems connected with analysis of the dynamics of short pulses in dispersive materials in the frequency domain.

1.8.1 Finite-difference time-domain method in direct integration of Maxwell equations

Let us examine a one-dimensional problem with field components E_x and H_y travelling in the z-direction. Supposing first for simplicity that the medium is isotropic, nonpermeable and nondispersive, we can write Maxwell's curl equations (1.44)–(1.45) in the form

$$\frac{\partial E_x}{\partial z} = -\mu_0 \frac{\partial H_y}{\partial t}, \tag{1.227}$$

$$\frac{\partial H_y}{\partial z} = -\frac{\partial D_x}{\partial t}. \tag{1.228}$$

Here the permeability μ_0 is that of free space, $D_x = \varepsilon E_x$, and the permittivity ε in this model is frequency-independent.

Transforming the differential equations (1.227)–(1.228) to the finite-difference form, we should consider the field Ψ not to be continuous but to be a kind of lattice of spacing h, where h is small but finite. The field Ψ will have meaning only at the net points (nodes), and the finite differences between the components for neighbouring points will be governed by difference equations. For the differential equations we mark off a rectangular net of spacing h in both directions and compute only at the net points, where the net lines cross. The value of Ψ at $x = mh, y = nh$ is then labeled $\Psi(m, n)$ or $\Psi^{(m)}(n)$.

There is a corresponding relation between the differential and difference equations. Related to the quantity $\Psi(n)$, which is a denumerable infinity of values Ψ for each integer value of n, is the continuous function $\Psi(x)$ of the continuous variable x. Related to the first, second etc. differences of $\Psi(n)$,

$$\left.\begin{aligned} \Delta\Psi(n) &= \Psi(n+1) - \Psi(n), \\ \Delta^2\Psi(n) &= \Psi(n+2) - 2\Psi(n+1) + \Psi(n) \end{aligned}\right\} \tag{1.229}$$

are the various differentials of $\Psi(x)$. Sometimes it is convenient to use instead of eqns (1.229) the central differentiation designated by the notation $\Psi(n \pm \frac{1}{2})$.

Now we can rewrite the curl equations (1.227)–(1.228) in finite-difference form:

$$H_y^{(n+\frac{1}{2})}\left(m+\frac{1}{2}\right) = H_y^{(n-\frac{1}{2})}\left(m+\frac{1}{2}\right) - \frac{\Delta t}{\mu_0 \Delta z}\left[E_x^{(n)}(m+1) - E_x^{(n)}(m)\right], \tag{1.230}$$

$$E_x^{(n+1)}(m) = E_x^{(n)}(m) - \frac{\Delta t}{\varepsilon \Delta z}\left[H_y^{(n+\frac{1}{2})}\left(m+\frac{1}{2}\right) - H_y^{(n+\frac{1}{2})}\left(m-\frac{1}{2}\right)\right]. \tag{1.231}$$

Here $\Psi^{(n)}(m)$ denotes the field components sampled at space point $z = m\Delta z$ and time point $t = n\Delta t$. Using the marching-in-time-and-space procedure, one can find the fields at a given space–time point, cheeking this numerical solution by means of appropriate numerical stability criteria (Tijhuis 1984). Space and time discretizations are selected to limit errors in the sampling process.

To improve this idealized model, the dispersion and absorption of the material should be taken into account. Let us consider this finite-difference time-domain (FDTD) method for solution of the time-domain Maxwell equations in Debye media.

Variations of electric fields in these media, characterized by finite relaxation times τ, are described by the function $\alpha(t)$, determined by the equation

$$\frac{d\alpha(t)}{dt} = -\frac{\alpha(t)}{\tau}, \tag{1.232}$$

The relaxation time τ may be temperature-dependent. The electric displacement D is defined as

$$D(t) = \varepsilon_\infty E(t) + \frac{\varepsilon_s - \varepsilon_\infty}{\tau} \int_{-\infty}^{t} \exp\left(-\frac{t-u}{\tau}\right) E(u)\,du. \tag{1.233}$$

Here ε_∞ is that of eqn (1.185). Differentiating eqn (1.233) with respect to t and making use of eqn (1.231), one can express the first derivative $\partial D/\partial t$ via the derivative $\partial E/\partial t$ (Frohlich 1958):

$$\frac{\partial D}{\partial t} = \varepsilon_\infty \frac{\partial E}{\partial t} + \frac{\varepsilon_s - \varepsilon_\infty}{\tau}\left[E(t) - \frac{\Delta t}{\tau} S(t)\right], \tag{1.234}$$

$$S(t) = \frac{1}{\Delta t} \int_{\infty}^{t} \exp\left(-\frac{t-u}{\tau}\right) E(\theta)\,du. \tag{1.235}$$

The function $S(t)$ can be reduced to the recursive form (Bui *et al.* 1991)

$$S(t) = S(t-\Delta t) \exp\left(-\frac{\Delta t}{\tau}\right) + \frac{1}{2}\left\{E(t-\Delta t) \exp\left(-\frac{\Delta t}{\tau}\right) + E(t)\right\}. \tag{1.236}$$

The second derivative $\partial^2 D/\partial t^2$ arising in the course of elimination of the function $H_y(t)$ from the system (1.227), (1.228) may be written as

$$\frac{\partial^2 D}{\partial t^2} = \varepsilon_\infty \frac{\partial^2 E}{\partial t^2} + \frac{\varepsilon_s - \varepsilon_\infty}{\tau}\left[\frac{E(t)}{\partial t} - \frac{E(t)}{\tau} + \frac{\Delta t}{\tau} S(t)\right]. \tag{1.237}$$

Replacing the derivatives of the electric field by finite differences, by analogy with eqn (1.230), we may obtain the second-order FDTD equation describing the values of components $E_x(z, t)$, $H_y(z, t)$ at the discrete space–time points. The same approach may be generalized for Lorentz media also (Joseph *et al.* 1991).

The accuracy of the results obtained by the FDTD method depends on the number of cells used for the computational discretization of the medium. In this connection, the results of time-domain computer modelling of reflected and transmitted pulses formed by normal incidence of a Hermitian-shaped Φ_1 eqn (1.30) transient on the surface of a collisional plasma layer have been compared with the envelopes of the reflected and transmitted pulses attained by means of steady-state complex reflection and transmission coefficients for each Fourier component of the incident pulse (Luebbers *et al.* 1991). The envelopes obtained by the two approaches were shown to agree closely if a plasma layer 1.5 cm thick (Langmuir frequency $\Omega_e = 1.8 \times 10^{11}\ \text{rad s}^{-1}$, relaxation time $\tau = 5 \times 10^{-11}$ s) was divided into 200 spatial cells each 75 mm thick, and a time step of 0.125 ps was used. The transient excitation of plasma 'ringing' close to the Langmuir resonant frequency, reminiscent of the similar effect in the neighbourhood of excitonic resonance in solids discussed above in Section 1.6, was also revealed by this computer simulation.

In general, the FDTD method allows modelling of arbitrarily shaped structures with various kinds of dispersion. On the other hand, one of the disadvantages of this method is that all the nodes of the mesh have to be calculated, regardless of whether the node is needed or not. Another disadvantage is associated with the enormous number of time steps needed to allow the observation of an extremely short pulse propagating in a medium with very high permittivity. However, alongside these computational problems associated with computer memory and the accuracy and stability of numerical algorithms, the interaction of short waveforms with dispersive media considered in the frequency domain is shown below to involve some problems of principle.

1.8.2 Problems with frequency-domain optics of ultrashort pulses

Fourier expansion is known to be a simple and powerful method for the investigation and design of slowly varying quasi-monochromatic wave fields in electromagnetic systems. However, the attempt to use this method formally for analysis of the interaction of transient fields with these systems is a matter of some difficulty. Thus the traditional picture of dispersive distortions of ultrashort pulses may provide an obvious demonstration of the merits and shortcomings of frequency-domain approach.

Let us consider a quasi-monochromatic plane wave with carrier frequency ω_0, incident from the air on the interface of a dispersive medium with refractive index $n(\omega)$. Supposing the bandwidth of this wave to be small compared with the carrier frequency, we can express the phase path of this wave φ at some position z inside the medium:

$$\varphi(\omega) = \frac{z \cdot \omega \cdot n(\omega)}{c}. \tag{1.238}$$

Expanding the function $\varphi(\omega)$ to a power series close to the carrier frequency ω_0,

$$\begin{gathered}\varphi(\omega) = \varphi(\omega_0) + \varphi'(\omega_0)\Omega + \tfrac{1}{2}\varphi''(\omega_0)\Omega^2 + \dots, \\ \Omega = \omega - \omega_0, \quad |\Omega| \ll \omega_0,\end{gathered} \tag{1.239}$$

and substituting the series (1.239) in the Fourier representation of the pulse envelope (Ginzburg 1967),

$$E(z,t)=\frac{1}{2\pi}\exp\left\{\mathrm{i}[\varphi(\omega_0)-\omega_0 t]\right\}\cdot\int_{-\infty}^{\infty}E_\Omega(\nu)\exp[\mathrm{i}\lambda(\eta,t,\Omega)]\,\mathrm{d}\eta\cdot\mathrm{d}\Omega, \tag{1.240}$$

$$\lambda(\eta,t,\Omega)=\Omega[\varphi'(\omega_0)+\eta-t]+\frac{\varphi''(\omega_0)\Omega^2}{2}+\frac{\varphi'''(\omega_0)\Omega^3}{6}+\ldots, \tag{1.241}$$

one can visualize the contribution of each term in eqn (1.241) to the evolution of an envelope $E(z,t)$. Thus, in the case of pulse propagation in an isotropic, lossless and homogeneous plasma with Langmuir frequency Ω_0 and refractive index $n_0=n(\omega_0)$,

$$n_0=\sqrt{(1-\Omega_e^2/\omega_0^2)} \tag{1.242}$$

the derivatives in the expansion (2.242) may be written as

$$\varphi'(\omega_0)=\frac{z}{cn_0},\quad \varphi''(\omega_0)=-\frac{z\cdot\Omega_e^2}{c\cdot\omega_0^3 n_0^3},\quad \varphi'''(\omega_0)=\frac{3z\cdot\Omega_e^2}{c\cdot\omega_0^4 n_0^5}. \tag{1.243}$$

Thus, restricting ourselves to the first term in expansion (1.241), we could consider the travel of the pulse with group velocity $v_g=cn_0$, its shape being conserved. Taking into account the next term in expansion (2.241), we face some mathematical difficulties connected with the dependence of integral (2.241) the initial pulse shape. In fact, this integral can be calculated analytically only for certain initial envelopes $E_0(t)|_{z=0}$.

The simplest model of amplitude-modulated pulse is the truncated sinusoid, eqn (1.8). Let us assume that the forward and backward edges of this rectangular pulse are incident on the boundary $z=0$ at the moments $t=-T$ and $t=T$ respectively. The dispersive spreading of this pulse in the depth of the plasma is described by well-known Fresnel integrals:

$$\left.\begin{aligned}E(z,t)&=\frac{1}{2}E_0(1+\mathrm{i})\exp\{\mathrm{i}[\varphi(\omega_0)-\omega_0 t]\}[F(u_1)-F(u_2)],\\ u_1&=\frac{1}{t_2}[T+t-\varphi'(\omega_0)],\quad u_2=\frac{1}{t_2}[T-t+\varphi'(\omega_0)],\end{aligned}\right\} \tag{1.244}$$

where t_2 is the characteristic time scale for the second-order dispersion:

$$t_2=\sqrt{[\pi|\varphi''(\omega_0)|]} \tag{1.245}$$

The dispersive spreading is accompanied by the lamination of the pulse envelope, depending on the correlation between the initial pulse duration $2T$ and the time scale t_2. Thus if the pulse is long enough ($2T\gg t_2$), the distortions of its central part are small, and the lamination proves to be localized close to the edges.

To improve this approximation for narrower pulses and increasing bandwidths, the cubic term in eqn (1.241) has been studied extensively. The influence of this third-order dispersion on the pulse diffusion may be examined only by numerical analysis, even for the above-mentioned simplest model of a rectangular-shaped pulse, eqn (1.8). Such analysis illustrates the loss of symmetry of some initially symmetrical pulses in the course of propagation (Wen and Chi 1989), characterized, unlike eqn (1.245), by the time scale t_3:

$$t_3 = \sqrt[3]{[|\varphi'''(\omega_0)|/2]}. \tag{1.246}$$

The pulse duration T does not influence the characteristic times (1.245)–(1.246), since the length of pulse path in the dispersive medium z is assumed to be much longer than the pulse scale cT. The corrections dictated by fourth-order dispersion, described by the Ω^4 term in eqn (1.241), have been discussed by Karlsson and Hook (1994).

To extend knowledge of the dynamics of ultrashort pulses, one might be tempted to continue this procedure, calculating the subsequent perturbations associated with the terms Ω^5, Ω^6 and so on in the phase expansion (1.241). However, the question of the convergence of such a perturbational procedure remains unsettled, and the influence of these corrections on the pulse evolution is unclear.

Hence attempts to apply the standard concept of travelling waves to the analysis of dispersive distortions of ultrashort broadband pulses face the problems inherent in the series of perturbations theories. It is expedient to stress three important points in regard to these attempts:

1. The coefficients of expansion (1.241) tend to infinity in the neighbourhood of the cutoff frequency of the medium.
2. The approach based on phase expansion to the power series (1.241) is useless for short pulses, when the spectral bandwidth becomes comparable with the carrier frequency.
3. In considering the refraction of waveforms without the carrier frequency, e.g. the Hermitian pulses (1.29), by means of Fourier expansions of their envelopes, both the aforesaid problems would be faced, especially when the time scale of the waveform is comparable with the period of the cutoff frequency $T_e = 2\pi\Omega_e^{-1}$. The dispersive degradation of a waveform develops rapidly in this case. Although the information about this process is encoded in the general Fourier representation, it can be interpreted only by numerical methods. However, one can lose a considerable amount of physical insight in the process of numerical solution.

To avoid these problems and to reconsider the optics of short waveforms in real dispersive and lossy media beyond the scope of the Fourier approach, a novel analytical theory is required. This theory will separate the physically meaningful phenomena in such optics from the numerical results valid for some given problem. Such non-Fourier optics is advanced below and a series of unforeseen phenomena, important for the optics of terahertz and femtosecond nonsinusoidal pulses, are discussed in Chapters 2 and 3.

1.9 Conclusions: modulated signals and incomplete Fourier expansions

We conclude with a few comments about the results that we have obtained in this Chapter.

1. The exact analytical solutions of equations of nonlinear geometric optics presented in Section 1.3, are expressed via the sums of eigenfunctions of hyperbolic equation (1.104). Such an approach is unusual in traditional mathematical physics: the solutions of hyperbolic equations are obtained mainly by integration along the characteristics of these equations, while the use of eigenfunctions is typical for elliptical equations, e.g. Laplace equation (1.84) (Morse and Feshbach 1953). However, one can recall the physically meaningful solution of elliptical equation (1.78) obtained in the case $\chi v_\omega < 0$ by integration along the characteristics in the complex (W, iv) plane (Lighthill 1967). Another example of the same approach, associated vice versa with the solution of hyperbolic equations by means of eigenfunctions, was demonstrated in Section 1.3. Both of these solutions are valid during the finite evolution time until the emergence of a breakdown of the initially smooth profile of the localized field.

2. The method of solution of the nonlinear Schrödinger equation used in Section 4.1 is reminiscent of the method of separation of variables in the theory of linear PDE: the unknown complex solutions U of eqn (1.130) are assumed to obey the condition of linear correlation between the real and imaginary parts of the function U. The coefficients of this correlation, being dependent on only one variable, are determined by the ODE. To check this solution, one should substitute it directly in the nonlinear Schrödinger equation. This method may be of interest for analysis of the series of non-stationary processes described by nonlinear PDE.

3. In discussing the interaction of modulated harmonic broadband pulses of radiation with a recording system, it should be pointed out that the same pulse can be detected either as a slowly varying modulated envelope or as one spectral line relevant to the monochromatic CW radiation, subject to the resolution properties of the detector. To exemplify this selectivity, let us examine the reception of amplitude-modulated oscillations with carrier frequency ω,

$$E = E_0[1 + k\cos(\Omega_m t)]\exp(-i\omega t), \tag{1.247}$$

by a resonant detector, considered as a contour with eigenfrequency ω_0 and attenuation decrement γ; for simplicity, the decrement γ is supposed to be frequency-independent. The equation describing the current I excited in the resonant contour by the signal (1.249) may be written as

$$\frac{\partial^2 I}{\partial t^2} + 2\gamma\frac{\partial I}{\partial t} + \omega_0^2 I = A_0[1 + k\cos(\Omega_m t)]\exp(-i\omega t). \tag{1.248}$$

Keeping in mind the resonant interaction $|\omega_0 - \omega| \ll \Omega_m$ and assuming the frequency of modulation Ω_m to be small compared with the carrier frequency, we can present the solution of (1.248) in the form:

$$\left.\begin{aligned} I &= \frac{iA_0}{2\omega_0\gamma}\left[1 + \frac{k\cos(\Omega_m t - \alpha)}{\sqrt{(1 + \Omega_m^2\gamma^{-2})}}\right]\exp(-i\omega t), \\ \operatorname{tg}\alpha &= \Omega_m\gamma^{-1}. \end{aligned}\right\} \tag{1.249}$$

Depending on the ratio $\Omega_m\gamma^{-1}$, one can outline two opposite situations, associated with 'slow' and 'fast' modulation with respect to the relaxation time of the contour γ^{-1}:

(a) 'Slow' modulation, high selectivity: $\Omega_m^2\gamma^{-2} \ll 1$. The signal in the contour (1.249) is described by a slowly varying current, proportional to the time-dependent spectral amplitude eqn (1.247).

(b) 'Fast' modulation, low selectivity: $\Omega_m^2\gamma^{-2} \gg 1$. Instead of modulated signal (1.247) the monochromatic sinusoid

$$I = \frac{iA_0}{2\omega_0\gamma}\exp(-i\omega t) \tag{1.250}$$

will be registered by the contour.

Thus such pulse phenomena are shown to be governed not by the ratio $\Omega_m\omega^{-1}$ characterizing the incoming modulated field, but by the ratio $\Omega_m\gamma^{-1}$ characterizing the receiver. In case of slow modulation, (a), the spectral amplitude of the recorded signal is oscillating in a quasi-stationary regime; in contrast, the fast modulation, (b), provides the recording of time-independent spectral amplitude. The variable spectral amplitude may be considered as an 'incomplete' Fourier expansion; its time-dependent part relates to the small frequencies ($\Omega_m^2 \ll \gamma^2$), which cannot be resolved by the receiver. In the choice of the time-domain or frequency-domain description of pulse interaction with a real system, its resolution properties should be taken into account.

2

TIME-DOMAIN ELECTRODYNAMICS OF CONTINUOUS MEDIA

2.1 Introduction: separable and nonseparable wave fields

This chapter is devoted to nonstationary propagation of nonsinusoidal electromagnetic fields in media with plasma-like dispersion, such as solid and gaseous plasmas, ionic crystals, and waveguide systems. The parameters of these media, widely used in optoelectronics and radiophysics, are considered here to be time-independent, whereas the propagating electromagnetic fields may possess significant nonstationary distortions due to dispersion and absorption. Unlike the usual approach associated with Fourier expansions of travelling plane waves and parameters of the medium, we shall operate directly, with time-domain Maxwell equations. To emphasize the distinctions between these approaches, let us outline some points of the traditional concept of quasi-monochromatic travelling waves. Using this concept, we limit our consideration to fields with the following properties:

1. The electric $\boldsymbol{E}$ and magnetic $\boldsymbol{H}$ components of the travelling electromagnetic field are described by sinusoidal oscillations with equal periods.
2. The wave train is long enough for its spectral bandwidth $\Delta\omega$ to be small compared with the carrier frequency ω.
3. The spatial–temporal structure of the EM field is represented as a product of coordinate-dependent and time-dependent functions. Thus the wave equation describing each component of the EM field $f(t, \boldsymbol{r})$,

$$\Delta f - \frac{1}{c^2}\frac{\partial^2 f}{\partial t^2} = 0, \tag{2.1}$$

is usually solved by means of a substitution,

$$f(t, \boldsymbol{r}) = \exp(-\mathrm{i}\omega t)\phi(\boldsymbol{r}), \tag{2.2}$$

where $\varphi(\boldsymbol{r})$ proves to be a solution of the Helmholtz equation. The solutions of eqn (2.1) in a form of product (2.2) arising by separation of variables belong to the class of so-called separable solutions. These solutions are used traditionally in the electrodynamics of harmonic waves.

However, it is significant that properties 1–3 above are not logical corollaries of the Maxwell equations. Neither the harmonic envelopes of EM fields nor their separable structure (2.2) are obligatory physical properties of these fields. Suppositions 1–3 are inherent in only one of the methods of solution of the Maxwell equations (1.45)–(1.48). Therefore it is expedient to analyse other solutions of these equations, free from limitations 1–3. It is noteworthy that, alongside stationary sinusoidal waves, wide classes of exact nonstationary analytical solutions of the Maxwell equations do exist. Unlike the traditional monochromatic wave trains, these solutions describe the nonsinusoidal rapidly deforming field envelopes in dispersive, guiding, magnetized and conductive media. Thus the proliferation of spatial–temporal structures of EM fields is not exhausted by separable solutions alone. The regimes of nonstationary propagation are described here by means of explicit nonseparable time-dependent solutions of the Klein–Gordon and telegraph equations, obtained beyond the scope of the Fourier approach.

To compare and contrast some features of sinusoidal and nonsinusoidal fields in the electrodynamics of continuous media, these features are derived here from the same equations, normalized by means of the natural spatial–temporal scales of the problems under discussion. Section 2.2 provides the mathematical background of time-domain EM fields in lossless media with plasma-like dispersion. This background is shown to be formed by exact nonstationary solutions of the Klein–Gordon equation, violating all the above-mentioned suppositions 1–3:

(1) the time scales of the oscillations of the electric and magnetic components given by these solutions differ;
(2) these time scales vary in the course of propagation;
(3) the rates of dispersive deformation of the field components are different.

Using these solutions, one can visualize some special properties of the rapid transformation of propagating EM fields, obtained first by sophisticated numerical analysis; thus the excitation of eigenfrequencies of the medium—'plasma ringing' (Luebbers *et al.* 1991)—appears automatically from these solutions (Subsection 2.2.2). The use of these regimes of propagation for the synthesis of time-dependent fields within such media is simplified essentially by recursive relations, reducing the differentiation and integration of nonseparable solutions to subtraction and summation of other solutions from this family. This reduction is promising for computational purposes.

Plasma-like dispersion covers a large class of physically meaningful materials and systems used in optics and microelectronics. This model is suitable for propagatlon of infrared and submillimetre radiation in semiconductors of $A^{III}B^{V}$ type with light free carriers, e.g. InSb or GaAs, when the frequency of collisions of carriers, $\nu \approx 0.5$ to $1 \times 10^{11}\,s^{-1}$ is much smaller than their Langmuir frequency $\Omega_e \approx 10^{13}\,rad\,s^{-1}$. The same model proves to be useful for the design of lenses, prisms and scintillation counters for IR radiation, produced from ionic crystals. These crystals, containing the ions at the nodes of the crystal lattice, such as NaCl or TiO_2, are distinguished by strong polarizability in the far IR and microwave regions. A useful area of application of this concept is associated with a wide range of electromagnetic fields propagating in gaseous plasmas. The energy transfer by nonseparable fields in such transparent dielectrics is examined in Section 2.3.

The generalization of time-domain electrodynamics for magnetized dielectrics with free carriers is advanced in Subsection 2.4.1. Similar regimes of anisotropic propagation of anharmonic fields in nonconducting ferrits are described by time-domain solutions of the Landau–Lifshitz equation. The spatial–temporal scales of nonseparable fields in media with spatial dispersion, defined in Subsection 2.4.2, allow the use of these normalized fields in the electrodynamics of such media also.

Section 2.5 is devoted to time-domain electrodynamics of conducting media. The exact nonstationary solutions of the telegraph equation represent explicit expressions for the electric and magnetic components of the field, which are not proportional to each other; the rapid deformations of the field in conducting media, being influenced by attenuation and spreading of the envelope, cannot be characterized, unlike harmonic fields, by attenuation decrement. The classical problem of electric current and voltage envelopes in transmission lines is reconsidered in Subsection 2.5.2 by reduction to the telegraph equation, and the relevant nonsinusoidal solutions of this equation are obtained.

The time-dependent modes in strip waveguides are analysed by means of nonseparable solutions of the Klein–Gordon equation in Section 2.6. The filtration property of these waveguides, filtering the fundamental mode of the projected nonsinusoidal radiation, is pointed out. The nonseparable structures of stationary monochromatic modes in microwave waveguides are examined.

Thus, in contrast to traditional Fourier electrodynamics of continuous media, the time-domain EM fields discussed in this chapter are

- nonsinusoidal
- nonstationary
- nonseparable.

A historical overview of the Klein–Gordon equation and its appllcations to electrodynamics and cross-disciplinary physics is given in Section 2.7.

2.2 The Klein–Gordon equation in the electrodynamics of dispersive media

This section is devoted to the mathematical background of the theory of nonstationary propagation of anharmonic electromagnetic fields in dispersive, lossless media. To develop this theory, free from the above-mentioned limitations of the Fourier approach, it is convenient to start from the classical problem of propagation of a plane, linearly polarized wave in a totally ionized, collisionless, isotropic electron plasma. This exactly solvable model proves to be useful for consideration of the wide classes of nonseparable electromagnetic fields in dispersive systems.

To analyse the dynamics of a plane wave with electric component E_x and magnetic component H_y, travelling in plasma in the z-direction, one can use the Maxwell equations (1.45)–(1.48). The electric displacement $\boldsymbol{D}$ in these equations is determined as

$$\boldsymbol{D} = \varepsilon_\infty \boldsymbol{E} + 4\pi \int_{-\infty}^{t} \boldsymbol{j}\,\mathrm{d}t. \tag{2.3}$$

Here ε_∞ is the high-frequency permittivity, resulting from dipole moments caused by the electronic and atomic polarizabilities, and $\boldsymbol{j}$ is the electric current induced in the medium by electric field $\boldsymbol{E}$. Neglecting the 'friction forces' acting on molecules and free carriers, we may determine the electron current $\boldsymbol{j}$ by the equation

$$\frac{\partial j}{\partial t} = \frac{\omega_\mathrm{p}^2 E}{4\pi} \tag{2.4}$$

where ω_p is the Langmuir plasma frequency (eqn 1.200).

To compare the stationary and nonstationary regimes of propagation of EM fields in a plasma, it is expedient to recall first the traditional Fourier approach to this problem. Thus, introducing the vector potential $\boldsymbol{A}$ and scalar potential φ,

$$\boldsymbol{E} = -\frac{1}{c}\frac{\partial \boldsymbol{A}}{\partial t} + \nabla\varphi, \quad \boldsymbol{H} = \operatorname{rot} \boldsymbol{A}, \tag{2.5}$$

one can derive from the basic equations (1.45)–(1.48) and (2.3) that the vector-potential component A_x for the linear field polarization $(A_y = A_z = 0)$ obeys the equation

$$\frac{\partial^2 A_x}{\partial z^2} - \frac{\varepsilon_\infty}{c^2}\frac{\partial^2 A_x}{\partial t^2} = \frac{\omega_\mathrm{p}^2}{c^2} A_x, \tag{2.6}$$

while the Lorentz condition

$$\operatorname{div} A + \frac{1}{c}\frac{\partial \varphi}{\partial t} = 0 \tag{2.7}$$

provides the value $\varphi = 0$ for this field. The same equation (2.7) is known to describe the field components E_x and H_y.

Making use of the travelling-wave concept and substituting the harmonic wave

$$A_x = A_0 \exp[\mathrm{i}(kz - \omega t)] \tag{2.8}$$

into eqn (2.6), we obtain the standard dispersive equation, relating the wavenumber k and frequency ω of a sinusoidal wave in an isotropic plasma:

$$k^2 v^2 = \omega^2 - \Omega_{\mathrm{e}}^2, \tag{2.9}$$

$$v = \frac{c}{\sqrt{\varepsilon_\infty}}, \quad \Omega_{\mathrm{e}}^2 = \frac{\omega_{\mathrm{p}}^2}{\varepsilon_\infty}. \tag{2.10}$$

Thus the equations of Maxwell electrodynamics (1.45)–(1.48) reduce in the case of free carriers (eqn 2.4) to eqn (2.6), which is known as the Klein–Gordon equation. Its stationary solution, characterized by time-independent amplitude and period, is represented by a harmonic wave train.

It is significant that, together with this standard solution (2.8), the Klein–Gordon equation has a multitude of nonseparable exact analytical solutions, describing the nonstationary fields. The explicit forms of these solutions are presented below.

2.2.1 Exact nonseparable solutions of the Klein–Gordon equation

We shall now reconsider the Klein–Gordon (KG) equation (2.6) without using the travelling-wave concept. Introducing the normalized variables η and τ and an unknown function f,

$$\eta = \frac{z\Omega_{\mathrm{e}}}{v}, \quad \tau = \Omega_{\mathrm{e}} t, \quad f(\tau, \eta) = A_x(\tau, \eta) \cdot A_0^{-1}, \tag{2.11}$$

we obtain from eqn (2.6) the following dimensionless form of the KG equation:

$$\frac{\partial^2 f}{\partial \eta^2} - \frac{\partial^2 f}{\partial \tau^2} = f. \tag{2.12}$$

Here Ω_{e} is the Langmuir resonant frequency (eqn 2.10); the normalization constant A_0 will be defined below.

Let us introduce the new variable

$$u=\sqrt{(\tau^2-\eta^2)}\cdot\operatorname{sign}\tau, \quad \operatorname{sign}\tau=\begin{cases}1, & \tau>0,\\ -1, & \tau<0,\end{cases} \tag{2.13}$$

where $\tau^2 \geqslant \eta^2$, and a new function $\psi_q(\tau,\eta)$ is

$$\psi_q(\tau,\eta)=\left(\frac{\tau-\eta}{\tau+\eta}\right)^{q/2} f_q(u), \tag{2.14}$$

the power q being unknown for the time being. Substitution of eqn (2.14) in the KG equation (2.12) yields the equation governing the function $f_q(u)$:

$$\frac{\partial^2 f_q}{\partial u^2}+\frac{1}{u}\frac{\partial f_q}{\partial u}+f_q-\frac{q^2 f_q}{u^2}=0. \tag{2.15}$$

The general solution of eqn (2.15) may be written as a sum of Bessel functions $J_q(u)$ and Neumann functions $N_q(u)$:

$$f_q(u)=a_q J_q(u)+c_q N_q(u), \tag{2.16}$$

where a_q and c_q are the arbitrary constants. Omitting the Neumann functions $N_q(u)$ tending to physically meaningless values for the fields $f_q(N_q\to\infty)$ at the points $u=0(\tau^2=\eta^2)$, we can write the solution $f(\tau,\eta)$ (2.14) in the form of a sum:

$$f(\tau,\eta)=\sum_q a_q\left(\frac{\tau-\eta}{\tau+\eta}\right)^{q/2} J_q(u). \tag{2.17}$$

This analysis leaves the value q as the sole unknown quantity in expression (2.14).

To define the parameter q one should compare the values of the function f (eqn 2.17) at a plane $\eta=0$[1], obtained in two ways:

1. It follows from eqn (2.17) that

$$f(\tau)\big|_{\eta=0}=\sum_q a_q J_q(\tau). \tag{2.18}$$

2. The general solution of hyperbolic equation (2.12) may be written as

$$f=C\cdot J_0(\sqrt{\tau^2-\eta^2})+\int_0^{(\tau+\eta)/2} U_1(t)\cdot J_0[\sqrt{(\tau-\eta)(\tau+\eta-2t)}]\cdot dt$$
$$+\int_0^{(\tau-\eta)/2} U_2(t)\cdot J_0[\sqrt{(\tau+\eta)(\tau-\eta+2t)}]\cdot dt. \tag{2.19}$$

[1] This plane $\eta=0$ may be considered as the boundary of the dispersive half-space (see Chapter 3).

Here J_0 is the Bessel function of zeroth order, C is a constant, and the arbitrary functions $U_1(t)$ and $U_2(t)$ determine the sources located at certain space–time points. The limits of integration,

$$x = \frac{\tau + \eta}{2}, \quad y = \frac{\eta - \tau}{2}, \tag{2.20}$$

are the characteristics of the hyperbolic equation (2.12). The first term in the sum (2.19) is the Green's function solution

$$f_a = J_0(u). \tag{2.21}$$

Considering the solution (2.19) at the interface of the dispersive medium $\eta = 0$ $(x = \tau/2,\ y = -\tau/2)$ and introducing the new variables

$$t = \frac{\tau}{2}u, \quad v^2 = \tau^2(1-u), \tag{2.22}$$

we obtain for example the field produced by the time-dependent source U_1 $(U_2 = 0,\ C = 0)$ in the form

$$f(\tau)\,|_{\eta=0} = \frac{1}{2}\int_0^{\tau} U_1\left[\frac{\tau}{2}\left(1 - \frac{v^2}{\tau^2}\right)\right] J_0(v)\,v\,\mathrm{d}v. \tag{2.23}$$

The wide classes of functions $U_1(t)$ may be described without loss of generality by means of analytical functions, represented by the power series without singularities

$$U_1(t) = \sum_{q=0}^{\infty} a_q t^q. \tag{2.24}$$

The arbitrary coefficients a_q provide the flexibility of representation of eqn (2.24) containing only integer nonnegative powers q. These physically meaningful distributions $U_1(t)$ are considered below. Substituting $U_1(t)$ (eqn 2.24) in eqn (2.23), we find the function $f(\tau)\,|_{\eta=0}$ analytically:

$$f(\tau)\,|_{\eta=0} = \sum_{q=1}^{\infty} a_q J_q(\tau). \tag{2.25}$$

The term with $q = 0$ in (2.25), relating to the time-independent part of the distribution U_1 at the boundary $\eta = 0$ is omitted in the solution (2.25).

Comparing now the solutions (2.25) and (2.18), obtained by different methods, one can conclude that the nonseparable solution of the KG equation (2.12) may be written in case (2.24) as a sum:

$$f(\tau, \eta) = \sum_{q=1}^{\infty} a_q \Psi_q, \quad q = 1, 2, 3, \ldots, \tag{2.26}$$

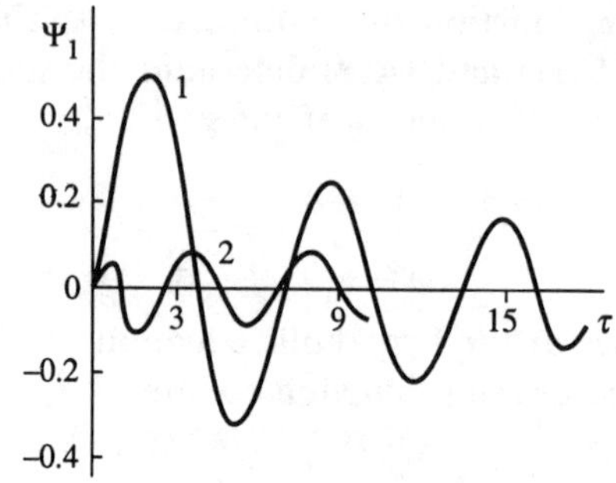

Fig. 2.1 Temporal evolution of nonsinusoidal harmonic Ψ_1 of the vector potential, eqn (2.27); the Ψ_1 envelopes are plotted vs. normalized time τ for $\eta = 0$ (curve 1) and $\eta = 10$ (curve 2).

$$\Psi_q(\tau, \eta) = \left(\frac{\tau - \eta}{\tau + \eta}\right)^{q/2} J_q(\sqrt{\tau^2 - \eta^2}). \tag{2.27}$$

The coefficients a_q are determined by the temporal variations of the electric and magnetic fields at the interface $\eta = 0$ (see Chapter 3).

The function (2.27) cannot be written as a product of time-dependent and coordinate-dependent factors. This function represents the simplest example of a nonseparable solution of the KG equation (2.12). The presence of Bessel function $J_q(u)$ in expression (2.27) provides the oscillations of this solution, such oscillations being nonsinusoidal.

Solution (2.26)–(2.27), defined for the values $\tau^2 \geqslant \eta^2$, describes the functions Ψ_q in both the regions $\tau > \eta \geqslant 0$ and $\tau \leqslant \eta \leqslant 0$. The function Ψ_q is symmetric (antisymmetric) with respect to the change $\tau \to -\tau$, $\eta \to -\eta$ subject to the even (odd) values of the order q. Therefore the subsequent examples illustrate, for simplicity, the behaviour of harmonics Ψ_q only in the half-space $\tau \geqslant \eta \geqslant 0$; to visualize the relevant envelopes in the region $\tau \leqslant \eta \leqslant 0$, the symmetry properties may be used.

The rapid diffusion of harmonics Ψ_q in the depth of a dispersive medium is shown in the Fig. 2.1; comparing curves 1 and 2 in this figure, one can see that the main maximum of the profile Ψ_1, arising at the interface of the dispersive medium ($\eta = 0$), is damped about fivefold, owing to propagation at a distance as small as $\eta = 10$. Curve 2 corresponds to the local time τ, at the position $\eta = \eta_1 > 0$. The time variables τ and τ_1 at the space–time points $(\tau, 0)$ and (τ_1, η_1) are related by

$$\tau_1 = \tau - \eta_1. \tag{2.28}$$

It is interesting to note that the solutions (2.27) might be obtained formally from the well-known solutions of the two-dimensional Helmholtz equation, written with polar coordinates (ρ, φ) on a plane (x, y):

$$\frac{\partial^2 f}{\partial\rho^2}+\frac{1}{\rho}\frac{\partial f}{\partial\rho}+\frac{1}{\rho^2}\frac{\partial^2\varphi}{\partial\varphi^2}+f=0, \tag{2.29}$$

$$\rho=\sqrt{(x^2+y^2)}, \quad \varphi=\operatorname{arctg}\frac{y}{x}, \tag{2.30}$$

$$f=\sum_{q=0}^{\infty} a_q\cdot J_q(\rho)\cdot\exp(\mathrm{i}q\varphi). \tag{2.31}$$

Using the substitution $x=\tau$, $y=\mathrm{i}\eta$, one can transform the Helmholtz equation to the KG equation (2.12), the variable ρ (2.29) and the factor $\exp(\mathrm{v}q\varphi)$ becoming the expressions

$$\rho=\sqrt{(\tau^2-\eta^2)}, \quad \exp(\mathrm{i}q\varphi)=\left(\frac{\tau-\eta}{\tau+\eta}\right)^{q/2}. \tag{2.32}$$

Finally, each partial solution in the sum (2.31) is reduced to eqn (2.17). The substitution $x=\tau$, $y=\mathrm{i}\eta$ transforms the separable solution (2.31) to nonseparable one (2.17); however, the spectrum of the values q is not defined by these transformations, and additional consideration, resulting for example in eqn (2.26), is required.

Thus, as well as the harmonic solution (2.9), the KG equation has the series of exact, oscillating, anharmonic, nonseparable solutions (2.26)–(2.27). These solutions will be shown to form the mathematical basis for the nonstationary electrodynamics of transparent media with a plasma-like dispersion law.

2.2.2 Nonstationary electromagnetic fields in plasma-like media

To find the electric and magnetic fields characterized by the nonseparable vector potential (2.26)–(2.27), one can use the relations (2.5). Normalizing these relations by means of eqn (2.11), we obtain the fields E_x and H_y, travelling in the z-direction:

$$\left.\begin{aligned} &E_x=E_0e_x, \quad H_y=H_0h_y, \quad E_0=\frac{A_0\Omega_\mathrm{e}}{c}, \quad H_0=\frac{A_0\Omega_\mathrm{e}}{c}\sqrt{\varepsilon_\infty}, \\ &e_x=-\frac{\partial f}{\partial\tau}, \quad h_y=\frac{\partial f}{\partial\eta}. \end{aligned}\right\} \tag{2.33}$$

The dimensionless electric (e_x) and magnetic (h_y) components are determined by substitution of the vector potential, eqns (2.26)–(2.27), in eqn (2.5). The differentiation of nonseparable solutions $\Psi_q(\tau, \eta)$ must be performed by important formulae based on the properties of Bessel functions (Appendix, eqn (A.2.1)):

$$\frac{\partial \Psi_q(\tau, \eta)}{\partial \tau} = \frac{1}{2}[\Psi_{q-1}(\tau, \eta) - \Psi_{q+1}(\tau, \eta)], \tag{2.34}$$

$$\frac{\partial \Psi_q(\tau, \eta)}{\partial \eta} = -\frac{1}{2}[\Psi_{q-1}(\tau, \eta) + \Psi_{q+1}(\tau, \eta)]. \tag{2.35}$$

Using the relations (2.34)–(2.35) we may represent the expressions for the electric and magnetic components of the field (2.33) within a plasma-like medium in the following form (Shvartsburg 1994):

$$e_x = \sum_{q=1}^{\infty} a_q e_q, \quad h_y = -\sum_{q=1}^{\infty} a_q h_q, \tag{2.36}$$

$$\left.\begin{matrix} e_q \\ h_q \end{matrix}\right\} = \frac{1}{2}\left(\frac{\tau-\eta}{\tau+\eta}\right)^{(q-1)/2}\left[J_{q-1}(u) \mp \frac{\tau-\eta}{\tau+\eta} J_{q+1}(u)\right]. \tag{2.37}$$

Formulae (2.36)–(2.37) give the exact nonseparable solution describing the propagation of an electromagnetic field in a dispersive medium beyond the scope of Fourier expansions. The coefficients a_q are determined by the boundary conditions at the interface $\eta = 0$ (see Chapter 3).

To visualize some basic features of nonseparable electromagnetic fields, it is convenient to examine first their time-dependent profiles at the interface of the dispersive medium ($\eta - 0$)

$$\left.\left.\begin{matrix} e_q(\tau) \\ h_q(\tau) \end{matrix}\right\}\right|_{\eta=0} = \frac{1}{2}[J_{q-1}(\tau) \mp J_{q+1}(\tau)]. \tag{2.38}$$

The profiles e_q and h_q relating to the values $q = 1$ and $q = 2$ are shown in the Figs 2.2 and 2.3. It is worth emphasing the following characteristics of these field envelopes:

1. The shape of the oscillations is not sinusoidal, and the intervals between two consecutive zeros of the envelopes are unequal.
2. The peak amplitudes of the oscillating distributions of the electric components as well as the magnetic components decrease at the wings of the distributions.
3. The rates of decrease of the electric and magnetic peak amplitudes differ, so their ratio is not constant even in a homogeneous medium.

Propagation of the nonseparable harmonics e_q and h_q (eqn 2.37) into the depth of the dispersive medium ($\eta > 0$) is accompanied by rapid diffusion of their envelopes. This process is illustrated by the damping of the first maximum of the travelling non-separable harmonic Ψ_1 (Fig. 2.4). The dynamics of an envelope of the field is exemplified by Fig. 2.5 for the electric

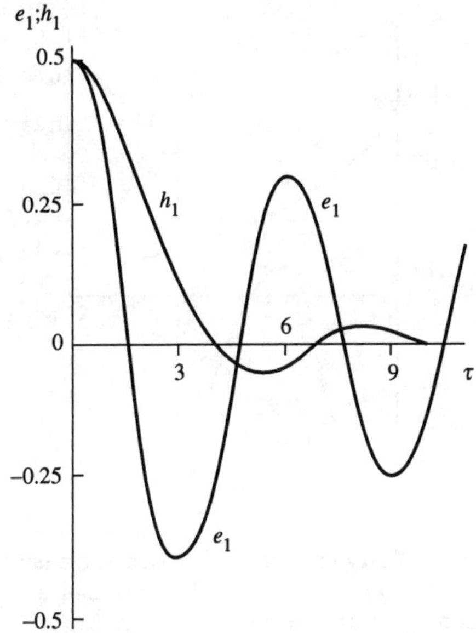

Fig. 2.2 Normalized envelopes of the electric e_1 and magnetic h_1 components of a nonsinusoidal field produced by harmonic Ψ_1 (eqn 2.37) at the interface $\eta = 0$, as a function of normalized time τ.

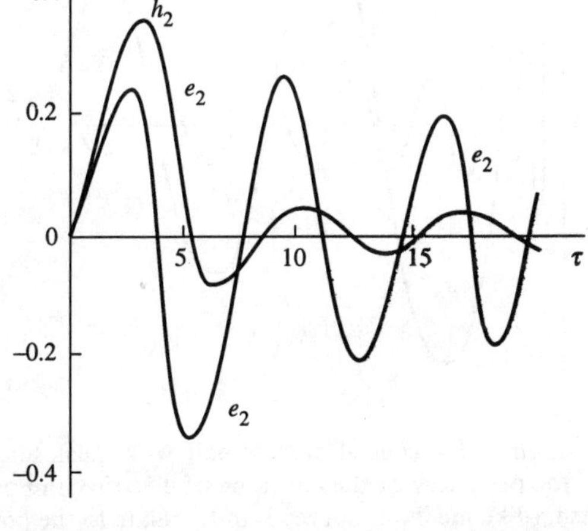

Fig. 2.3 Normalized envelopes of the electric e_2 and magnetic h_2 components of a nonsinusoidal field at the interface $\eta = 0$, vs. normalized time τ.

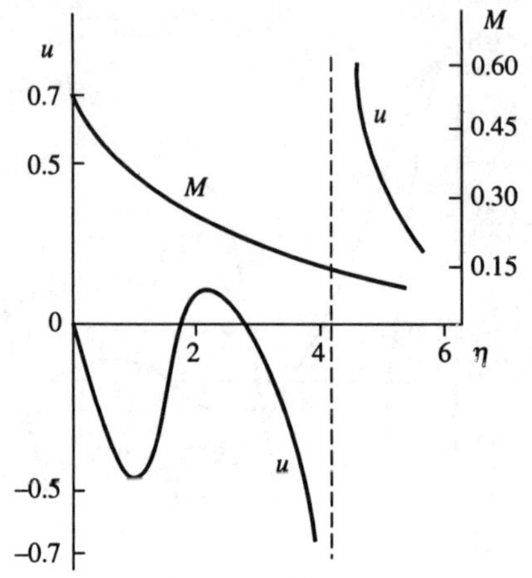

Fig. 2.4 Diffusion of the first maximum of nonseparable harmonic Ψ_1 (eqn 2.27). The peak amplitude of this maximum M and its normalized velocity u (eqn 2.72) are plotted vs. position η within the medium.

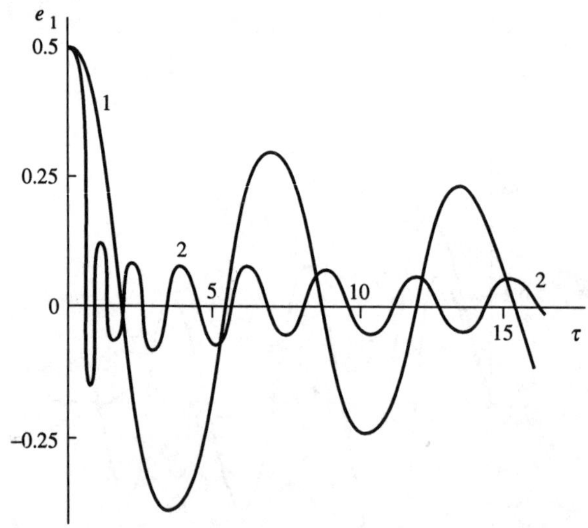

Fig. 2.5 Conservation of the central narrow half-wave peak and quasi-harmonic transformation of the periphery of the envelope of electric component e_1 travelling in a dispersive plasma-like medium; curves 1 and 2 relate to the positions $\eta = 0$ and $\eta = 50$ respectively, τ is the normalized time.

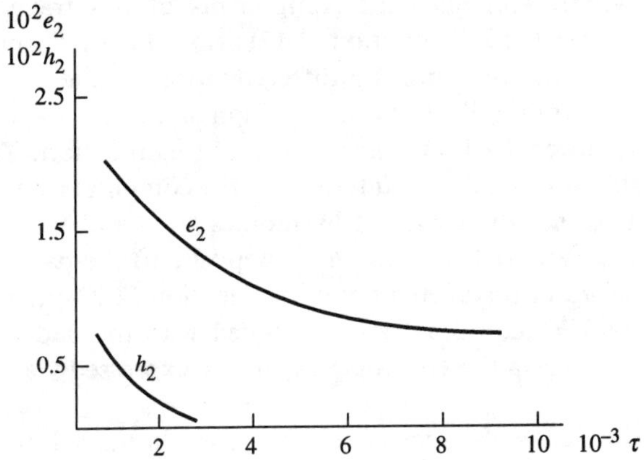

Fig. 2.6 Diffusing envelopes of the electric e_2 and magnetic h_2 harmonics in the depth of a plasma-like medium ($\eta = 50$).

harmonic e_1, since the relevant harmonic of the magnetic component is damped faster at the wings ($\tau^2 \gg \eta^2$) owing to dispersion (Fig. 2.6). In considering these figures, the relation (2.28), defining the time variable in the depth of the medium, must be taken into account. The harmonic e_1, as well as h_1, is distinguished by the invariance of the main peak $e_1 = h_1 = 0.5$, located at the space–time point $\tau = \eta$, in the course of propagation (Fig. 2.5); in contrast, all the other envelopes are characterized by the values $e_q = 0$ and $h_q = 0$ $(q \neq 1)$ at the same point $\tau = \eta$).

The dispersive diffusion of the magnetic components h_q at the wings of their envelopes develops faster than that of the electric components e_q. These processes may be examined by substitution of the asymptotes of the Bessel functions $J_q(u)$ in the range $\tau \gg \eta$, $u = \tau \gg 1$ (eqn A.2.2) in eqn (2.37):

$$e_q = \sqrt{2/(\pi\tau)} \cdot \cos\left(\tau - \frac{\pi q}{2} + \frac{\pi}{4}\right), \quad h_q = \frac{\eta}{\tau} e_q. \tag{2.39}$$

The terms of order of magnitude $\eta\tau^{-1} \ll 1$ are omitted in the expression for e_q eqn (2.39). Thus the evolution of Bessel fields in the depth of the medium ($\eta \gg 1$) results in the formation of harmonic oscillations at the wings of the envelopes e_q and h_q. The frequency of these oscillations is close to the cutoff frequency of the medium Ω_e. The amplitudes of the magnetic oscillations are smaller than those of the electric ones (eqn 2.39); this tendency is shown in Fig. 2.6.

Both the electric and magnetic components of this travelling electromagnetic field obey the KG equation (2.12) also. The qth harmonic of each of these components obtained by differentiation of the vector-potential function may be found algebraically through summation and subtraction of the $(q-1)$th and $(q+1)$th harmonics of this function. This property of nonseparable solutions is of interest for the computational purposes. It is useful to note that the parity of harmonics Ψ_{q-1} and Ψ_{q+1} forming the electric and magnetic fields e_q and h_q is opposite to the parity of the relevant harmonic Ψ_q in the vector-potential function (2.27); hence, the even envelopes e_1 and h_1 (eqn 2.37) are associated with the odd harmonics Ψ_0 and Ψ_2, and vice versa the envelopes e_2, h_2 are expressed via the functions Ψ_1 and Ψ_3.

It is noteworthy that by regrouping the terms in the solution (2.26), we can represent this solution in the form

$$f=\sum_{q=2}^{\infty} b_q\theta_q(\tau,\eta), \quad \theta_q=\frac{1}{2}[\Psi_{q-1}-\Psi_{q+1}]. \tag{2.40}$$

Then the dimensionless electric and magnetic fields may be written, by analogy with eqn (2.36), as

$$e_x=-\sum_{q=2}^{\infty} b_q e'_q, \quad h_y=-\sum_{q=2}^{\infty} b_q h'_q, \tag{2.41}$$

$$e'_q=\frac{1}{4}[\Psi_{q-2}-2\Psi_q+\Psi_{q+2}], \tag{2.42}$$

$$h'_q=\frac{1}{4}[\Psi_{q-2}-\Psi_{q+2}]. \tag{2.43}$$

The expressions (2.41)–(2.43) and (2.36)–(2.37), containing the same functions Ψ_q, are equivalent; the coefficients b_q in eqn (2.40) may be expressed in terms of the coefficients a_q in eqn (2.36), e.g.

$$\left.\begin{aligned} b_2=2a_1,\quad b_4=2(a_1+a_3),\quad b_6=2(a_1+a_3+a_5),\\ b_3=2a_2,\quad b_5=2(a_2+a_4),\quad b_7=2(a_2+a_4+a_6).\end{aligned}\right\} \tag{2.44}$$

Since the higher derivatives of the functions Ψ_q also obey the KG equation, other representations of the solution of this equation based on such derivatives, similar to eqn (2.42), may be used. The selection of a suitable representation is determined by the problem discussed. Thus the solution (2.36)–(2.37) is convenient for consideration of the components of the electromagnetic field; to find the integral transformations of these fields, e.g. the electric displacement (eqn 2.3), the presentation (2.41)–(2.43) proves to be suitable, since the integration can be performed by means of the formula

$$\int [\Psi_{q-2}(\tau, \eta) - 2\Psi_q(\tau, \eta) + \Psi_{q+2}(\tau, \eta)]\, d\tau = 2[\Psi_{q-1}(\tau, \eta) - \Psi_{q+1}(\tau, \eta)]. \tag{2.45}$$

Together with the differentiation of the nonseparable functions $\Psi_q(t, \eta)$, the integration of these functions can be reduced to simple algebraic summation and subtraction.

Thus, without the addition of any new terms to the traditional Maxwell equations (1.45)–(1.48), exact nonseparable solutions of these equations are shown to exist as well as the usual sinusoidal waves. The above-mentioned characteristics 1–3 of such anharmonic Bessel-like nonstationary solutions violate the traditional assumptions (see Section 2.1). These solutions, obtained without the use of Fourier expansions, form the analytical basis of non-Fourier electrodynamics.

2.3 Bessel fields in transparent dielectrics

The dynamical electromagnetic properties of dielectrics and semiconductors attract attention in the design of numerous optoelectronic systems based on the rapid polarization and absorption of these materials in the infrared region. Such processes induced by external alternating electric fields cannot respond to these fields instantaneously. The time delay in the relaxation of polarization is influenced by the masses of molecular dipoles and charge carriers and by the interaction forces impeding their movement. The coupled variations of polarization $\boldsymbol{P}$ and electric field $\boldsymbol{E}$ in isotropic dielectrics with free carriers (semiconductors, gaseous plasmas, ionic solids) are governed by the pair of equations (Zeeger 1973)

$$\frac{\partial^2 \boldsymbol{P}}{\partial t^2} + \gamma \frac{\partial \boldsymbol{P}}{\partial t} + \omega_0^2 \boldsymbol{P} = \frac{\omega_2^2}{4\pi} \boldsymbol{E}, \tag{2.46}$$

$$\frac{\partial^2 \boldsymbol{E}}{\partial z^2} - \frac{\varepsilon_\infty}{c^2} \frac{\partial^2 \boldsymbol{E}}{\partial t^2} = \frac{4\pi}{c^2} \frac{\partial \boldsymbol{j}}{\partial t}. \tag{2.47}$$

Parameters γ and ω_0 characterize the attenuation and eigenfrequency of polarization oscillations, ω_p is the Langmuir frequency (1.200), and ε_∞ is the high-frequency dielectric permittivity. The values of ε_∞ for the semiconductors and ionic crystals used in optoelectronics are ~12–16; $\varepsilon_\infty = 1$ for gaseous plasmas. The eigenfrequency ω_0 and plasma frequency ω_p are usually in the IR region.

When the time scales of variations in the electric field T are long enough ($T\omega_0 \gg 1$) and the attenuation is small ($\gamma \ll \omega_0$), the polarization $\boldsymbol{P}$ proves to be proportional to the field $\boldsymbol{E}$ eqn (2.46), the envelope $\boldsymbol{E}$ being arbitrary:

$$P = \frac{\omega_p^2}{4\pi\omega_0^2} E. \tag{2.48}$$

Substituting eqn (2.48) in eqn (2.47), one could consider the nondispersive model of the medium, providing distortion-free propagation of the signal E with a velocity

$$v_0 = \frac{c}{\sqrt{(\varepsilon_\infty + \omega_p^2/\omega_0^2)}}. \tag{2.49}$$

In contrast to this quasi-stationary model, attention will be focused below on a different case, associated with the ultrafast variations of polarization when the eigenfrequencies ω_0 may be neglected ($\omega_0 T \ll 1$). Introducing the electric current j associated with the temporal variations of polarization,

$$j = \frac{\partial P}{\partial t}, \tag{2.50}$$

we reduce eqn (2.46) to the simplified form

$$\frac{\partial j}{\partial t} + \nu j = \frac{\omega_p^2 E}{4\pi}, \tag{2.51}$$

related to the Debye model of dielectrics (Debye 1929). Neither the polarization P nor the electric displacement D in this nonstationary model is proportional to the electric field E. The definition (2.3) must be consldered in this case as an integral operator connecting the functions D and E. Nevertheless, the electric displacement for the nonseparable Bessel fields may be found in the time domain analytically. This approach permits the concepts of energy density and power flux (Poynting vector) to be generalized for these nonstationary fields.

2.3.1 Time-domain electrlc displacement induced by Bessel fields

The electric displacement is known to be an important characteristic of a dielectric medium, widely used in the analysis of its power-transfer properties, such as the energy density or group velocity of the field. This characteristic is usually considered for the stationary oscillations of molecular dipoles and free carriers excited by long trains of sinusoidal waves. Such stationary processes are described traditionally in the frequency domain, the spectral amplitudes of electric displacement D_ω and electric field E_ω being proportional [$D_\omega = \varepsilon(\omega) \cdot E(\omega)$]; the coefficient of proportionality $\varepsilon(\omega)$ is the dielectric permittivity. The above-mentioned model of high-frequency oscillations ($\omega^2 \gg \omega_0^2$) in a lossless medium ($\omega \gg \gamma$) is characterized by a permittivity

$$\varepsilon_\omega = \varepsilon_\infty \left(1 - \frac{\Omega_e^2}{\omega^2}\right). \tag{2.52}$$

The concept of dielectric permittivity becomes invalid for rapidly varying nonsinusoidal fields; however, the nonstationary electric displacement may be analysed for such fields without using this concept. To examine the nonstationary electric displacement $D(z, t)$ for nonseparable fields in the time domain, one should use the general definition (2.3). To reveal the physically meaningful features of this approach, let us consider first the model (2.51) for the case of weak dissipation ($\nu \ll \Omega_0$). Ignoring the losses to a first approximation and making use of eqns (2.46), (2.50) and (2.51), we obtain the electric field $E(z, t)$, obeying the KG equation, in the form (2.41)–(2.42). Substitution of these Bessel-like harmonics in eqn (2.51) yields the electric current induced by the field $E(z, t)$:

$$j(\tau, \eta) = -\frac{\varepsilon_\infty \Omega_e^2 A_0}{4\pi c} \sum_{q=2}^{\infty} \frac{a_q}{2} [\Psi_{q-1}(\tau, \eta) - \Psi_{q+1}(\tau, \eta)]. \tag{2.53}$$

Substituting eqn (2.53) in eqn (2.3), one can find the electric displacement $D(z, t)$ as a sum:

$$D(z, t) = -\frac{\varepsilon_\infty \Omega_e^2 A_0}{c} d(\tau, \eta), \quad d(\tau, \eta) = \sum_{q=2}^{\infty} b_q d_q, \tag{2.54}$$

$$d_q(\tau, \eta) = \frac{1}{4} [\Psi_{q-2}(\tau, \eta) + 2\Psi_q(\tau, \eta) + \Psi_{q+2}(\tau, \eta)]. \tag{2.55}$$

The behaviour of harmonics e_3 and d_3 in the time domain is shown in Fig. 2.7. Both e_q and d_q harmonics spreading rapidly in the course of propagation, and the envelopes e_q and d_q at any given position prove to be very different.

It is worth outlining again the equivalence of expressions (2.37) and (2.42) for the nonseparable fields. Thus, using eqn (2.34), one could express the dimensionless electric displacement $d(\tau, \eta)$ in the form:

$$\left.\begin{aligned} d(\tau, \eta) &= \sum_{q=2}^{\infty} a_q d_q(\tau, \eta), \\ d_q(\tau, \eta) &= \frac{1}{2} \left[\Psi_{q-1}(\tau, \eta) - \Psi_{q+1}(\tau, \eta) + 2\int \Psi_q(\tau, \eta)\, d\tau\right]. \end{aligned}\right\} \tag{2.56}$$

Calculating the integral in eqn (2.56) by means of a recursive formula following from eqn (2.34),

$$\int \Psi_q(\tau, \eta)\, d\tau = 2 \sum_{p=1}^{\infty} \Psi_{q+2p-1}, \tag{2.57}$$

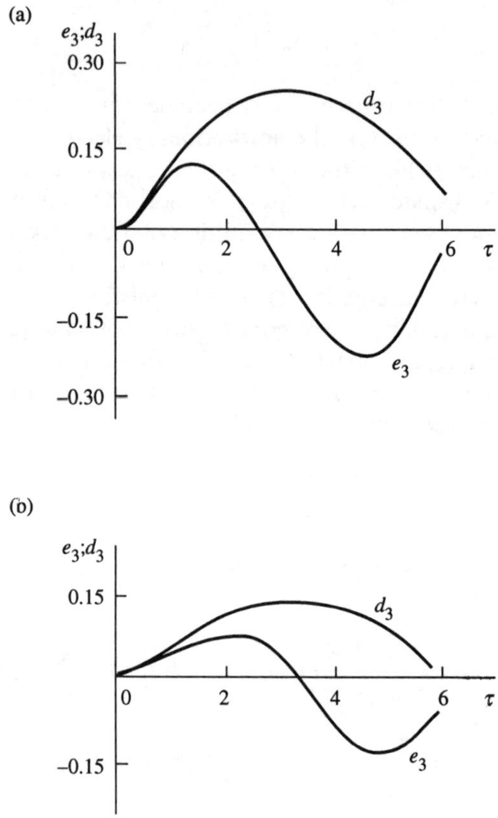

Fig. 2.7 Dispersive diffusion of harmonics of electric fluid e_3 (eqn 2.42) and electric displacement d_3 (eqn 2.53) at the positions (a) $\eta = 0$ and (b) $\eta = 1$, vs. normalized time τ.

and keeping in mind the relations (2.44) between the coefficients a_q and b_q, we can be convinced of the identity of solutions (2.37) and (2.42). To visualize the contribution of each harmonic d_q to the electric displacement D, using the finite sums and avoiding the infinite series (2.57), the expression (2.54) is more suitable for the problem discussed.

The contribution of the eigenfrequencies ω_0 to the electric displacement was neglected above. To take into account small but finite values of the ratio $\omega_0^2 \cdot \Omega_e^{-2} \ll 1$, the theory of perturbations for nonseparable fields can be developed. Let us present the polarization $\boldsymbol{P}$ and the current $\boldsymbol{J}$ in the forms $\boldsymbol{P} = \boldsymbol{P}_0 + \boldsymbol{P}_1$, $\boldsymbol{J} = \boldsymbol{J}_0 + \boldsymbol{J}_1$, where the unperturbed and perturbed values are labelled with the subscripts 0 and 1 respectively. The current $\boldsymbol{j}_1$ is governed by the equation

$$\frac{\partial \boldsymbol{J}_1}{\partial t} = -\omega_0^2 \cdot \boldsymbol{P}_0, \tag{2.58}$$

following from eqn (2.46) in the lossless model ($\nu = 0$). Determining the unperturbed functions $\boldsymbol{P}_0$ and $\boldsymbol{J}_0$ from the solution of the KG equation, chosen in the form

$$f = \frac{1}{8} \sum_{q=3}^{\infty} b_q [\Psi_{q-3} - 3 \cdot \Psi_{q-1} + 3 \cdot \Psi_{q+1} - \Psi_{q+3}], \tag{2.59}$$

we can find the current perturbation from $\boldsymbol{j}_1$ from eqn (2.58). Substitution of $\boldsymbol{J}_0$ and $\boldsymbol{J}_1$ in eqn (2.3) results in the following formula for nonseparable harmonics in a representation similar to eqn (2.54):

$$d_q = \frac{1}{16} [\Psi_{q-4} - 2 \cdot \Psi_q + \Psi_{q+4}] + \frac{\omega_0^2}{\Omega_e^2} \Psi_q. \tag{2.60}$$

The summation in eqn (2.54) in this case is begun from the value $q = 4$. The last term in eqn (2.60) describes the contribution of the eigenfrequencies ω_0 to the time-domain electric displacement.

Thus the electric displacement for nonseparable fields is introduced here in the time domain directly by means of the general definition (2.3). The values of electric displacement $\boldsymbol{D}(z, t)$ and electric field $\boldsymbol{E}(z, t)$ show that the temporal variations of these fields at any position in the medium are not proportional to each other. The function $\boldsymbol{D}(z, t)$ describes the physically meaningful characteristic of a nonstationary and spatially heterogeneous EM field in a stationary and homogeneous dielectric medium beyond the scope of the dielectric permittivity concept. Here the energy-carrying characteristics of this medium, such as the energy density or group velocity, depending on the electric displacement, are introduced below for Bessel-like fields in the time domain again without using the dielectric permittivity function.

2.3.2 Energy flux of a nonsinusoidal electromagnetic field in a semiconductor

The energy flux in a lossless medium, described by the Poynting vector, can be written in the general form

$$\boldsymbol{P} = \frac{c}{4\pi} [\boldsymbol{E} \cdot \boldsymbol{H}]. \tag{2.61}$$

Using the density of electromagnetic energy W, expressed through the electric displacement $\boldsymbol{D}$ (Ginzburg 1967),

$$W = \frac{1}{4\pi} \int \boldsymbol{E} \cdot \frac{\partial \boldsymbol{D}}{\partial \tau} \mathrm{d}\tau + \frac{|\boldsymbol{H}|^2}{8\pi}, \tag{2.62}$$

one can introduce the group velocity of the field,

$$v_g = \frac{P}{W}. \tag{2.63}$$

Taking into account the equation

$$\frac{\partial D}{\partial \tau} = -\frac{\Omega_e A_0}{c}\left(\frac{\partial^2 f}{\partial \tau^2} + f\right) \tag{2.64}$$

following from eqn (2.3), we obtain an explicit expression for the energy density W, depending upon the dimensionless function f and its derlvatives (eqn 2.33):

$$W = \frac{\Omega_e^2 A_0^2}{8\pi c^2}\left[|f|^2 + \left|\frac{\partial f}{\partial \tau}\right|^2 + \left|\frac{\partial f}{\partial \eta}\right|^2\right]. \tag{2.65}$$

The energy flux (eqn 2.61) may be written by means of the same function f

$$P_z = -\frac{\Omega_e^2 A_0^2}{4\pi c}\frac{\partial f}{\partial \tau}\cdot\frac{\partial f^*}{\partial \eta}. \tag{2.66}$$

Substitution of eqns (2.65)–(2.66) in the definition (2.63) yields the expression for the group velocity

$$v_g = -2\cdot c\cdot\frac{\partial f}{\partial \tau}\cdot\frac{\partial f^*}{\partial \eta}\cdot\left[|f|^2 + \left|\frac{\partial f}{\partial \tau}\right|^2 + \left|\frac{\partial f}{\partial \eta}\right|^2\right]^{-2}. \tag{2.67}$$

Formula (2.67) is valid for both stationary and nonstationary electromagnetic fields. Thus, in particular, the group velocity of stationary sinusoidal waves may be obtained by substitution of the harmonic vector-potential function $f = \exp[i(kz - \omega t)]$, describing these waves in eqn (2.67):

$$v_g = \frac{c^2 k}{\omega \varepsilon_\infty}. \tag{2.68}$$

The same result follows from the traditional formula derived for sinusoidal waves,

$$\bar{v}_g = \frac{\partial \omega}{\partial \bar{k}}, \tag{2.69}$$

when the frequency ω and the wavenumber k are connected via the dispersive relation (2.9). However, compared with eqn (2.69), formula (2.67) is much more general, since it may be used for analysis of group velocities of both stationary wave trains and nonstationary anharmonic fields when the frequency ω and wavenumber k are not defined:

$$\frac{v_g}{c} = -\frac{2\dfrac{\partial f}{\partial \tau}\cdot\dfrac{\partial f^*}{\partial \eta}}{f^2 + \left(\dfrac{\partial f}{\partial \tau}\right)^2 + \left(\dfrac{\partial f}{\partial \eta}\right)^2} \tag{2.70}$$

Using the unequality $-1 \leqslant 2ab(a^2+b^2)^{-1} \leqslant 1$, and keeping in mind that the denominator in the fraction (2.70) is always positive, one can be convinced that the value v_g is located within the limits $-c \leqslant v_g \leqslant c$. The positive and negative values of the group velocity relate to the inflow and outflow of the energy respectively. The group velocity of Bessel-like fields, depending upon the position (eqn 2.70), may be characterized by temporal oscillations.

Unlike the constant group velocity of harmonic waves in a homogeneous stationary medium, the value v_g for the nonseparable fields in the same medium decreases in the far zone ($\tau^2 \gg \eta^2$); using the asymptotes (2.39), one can find

$$\left.\frac{v_g}{c}\right|_{\tau^2 \gg \eta^2} = \frac{\text{const}\cdot\eta}{\tau}. \tag{2.71}$$

Hence the power flux of the nonseparable field in a homogeneous medium is located near the interface $\eta = 0$.

Together with the group velocity, the propagation of harmonic waves in a transparent dielectric is characterized by the velocity of any fixed point at the wave profile: the so-called phase velocity $v_p = \omega k^{-1}$. In contrast, the attempt to define a similar velocity for the nonseparable fields is impeded because of the rapid dispersive distortions of the envelopes of these fields in the course of propagation. Thus, considering the movement of any maximum of the harmonic Ψ_q, one can determine the velocity of this maximum v_m via the derivatives of the envelope Ψ_q:

$$\frac{v_m}{c} = -\frac{\dfrac{\partial \psi_q}{\partial \tau}}{\dfrac{\partial \psi_q}{\partial \eta}}. \tag{2.72}$$

Substitution of the harmonic vector-potential function $\psi_q \rightarrow \exp[i(kz-\omega t)]$ in eqn (2.72) results in the usual expression for the velocity $u = \omega\cdot k^{-1}$, coinciding with the phase velocity of this harmonic wave. In contrast, consideration of e.g. the first maximum of the harmonic ψ_1 illustrates the variations in its velocity v_m (Fig. 2.4) and even the change in its direction; moreover, these variations are accompanied by a decrease in the maximum (Fig. 2.4, curve M). Such pictures are different for different maxima and different harmonics. Hence any attempt to introduce a phase velocity for these nonseparable fields would be useless[1].

Thus the concepts of time-dependent electric displacement $\boldsymbol{D}$, electromagnetic energy density W and power flux $\boldsymbol{P}$ are developed in the electrodynamics of nonseparable fields in transparent dielectrics and semiconductors without using Fourier representations. The following concepts, inherent in Fourier optics, are omitted in the time-domain approach.

(1) the frequency-dependent dielectric permitivity $\varepsilon(\omega)$;

(2) the phase of oscillations of the EM field and the phase velocity v_{p};

(3) the refractive index n, related to the phase velocity v_{p} $(n = cv_{\mathrm{p}}^{-1})$ as well as to the dielectric permittivity $(n = \varepsilon^{1/2})$.

The consideration below of nonseparable EM fields in the electrodynamics of magnetized and lossy media also proves to be free from these frequency-domain concepts.

2.4 Nonseparable fields in the optics of magnetized and spatially dispersive media

Analysis of propagation of electromagnetic fields in magnetized and crystal media is complicated by the anisotropy and spatial dispersion of these media. To visualize the influence of these phenomena on propagation of nonseparable fields, discussion will be given below to two limiting cases, relevant to longitudinal and transverse propagation of these fields with respect to an external magnetic field $\boldsymbol{H}_0$. These cases are examined for dielectrics with free carriers, e.g. plasmas, and nonconducting materials, e.g. magnetodielectrics.

Nonseparable fields in magnetoplasma may be considered by means of equations describing the current of free carriers $\boldsymbol{J}$. Let us introduce the coordinate system with the z-axis directed along the magnetic field $\boldsymbol{H}_0$ and the (x, y) plane orthogonal to the z-direction. Restricting ourselves for simplicity to the model of a collisionless electron plasma, we can write these equations in the form

---

[1] The sole exception is associated with the movement of the main maxima of haxmonics e_1 and h_1 eqns (2.34)–(2.35); these maxima, described by the Bessel function $J_0(u)$, remain constant in the course of propagation ($\tau = \eta$, $J_0(0) = 1$) as well as their velocities $v = c \cdot \varepsilon_\infty^{-1/2}$. This situation will be discussed in detail in Section 3.2.

$$\left.\begin{aligned}\frac{\partial j_x}{\partial t} &= \frac{\omega_{\mathrm{p}}^2}{4\pi}E_x + \omega_H j_y,\\ \frac{\partial j_y}{\partial t} &= -\omega_H j_x + \frac{\omega_{\mathrm{p}}^2}{4\pi}E_y,\\ \frac{\partial j_z}{\partial t} &= \frac{\omega_{\mathrm{p}}^2}{4\pi}E_z.\end{aligned}\right\} \tag{2.73}$$

Here ω_{p} is the plasma frequency (eqn 1.200) and ω_H is the gyrofrequency of free carriers with effective mass m^*:

$$\omega_{\mathrm{H}} = \frac{eH_0}{m^*c}. \tag{2.74}$$

Light carriers are prevalent in the semiconductors widely used in optical devices. Thus the value of m^* for n-type InSb is $m^* = 0.013m_0$, where m_0 is the free electron rest mass (Zeeger 1973). Herein the frequency ω_H is $\sim 8\times 10^{11}\,\mathrm{rad\,s^{-1}}$, the magnetic field H_0 being $\sim 0.1\,\mathrm{T}$. The frequency of electron collisions with the crystal lattice ν may be neglected in the system (2.73) in the low-temperature region, when the condition $\nu^2 \ll \omega_H^2$ is fulfilled ($T \leqslant 10\,\mathrm{K}$, $\nu \approx 10^{11}\,\mathrm{s^{-1}}$).

Unlike eqn (2.73), the electric current in ferromagnetic media without free carriers is determined by the magnetic moment per unit volume $\boldsymbol{M}$; the magnetic inductance of such media also depends on $\boldsymbol{M}$:

$$\boldsymbol{J} = c\cdot \operatorname{rot}\boldsymbol{M}, \quad \boldsymbol{B} = \boldsymbol{H} + 4\pi\boldsymbol{M}. \tag{2.75}$$

Oscillations of magnetic moment $\boldsymbol{M}$ are known to support the propagation of EM fields in magnetized dielectric media (Kittel 1948). Rapidly varying nonsinusoidal EM fields in media with polarization currents of different physical nature, eqns (2.73) and (2.75), are considered in Subsection 2.4.1.

Unlike these vortical EM fields, the propagation of harmonic curl-less oscillations of electric fields with frequencies close to the resonant frequencies of the medium is known to depend essentially on the spatial dispersion. The propagation of curl-less nonsinusoidal electric fields is also influenced by this effect. An important mechanism of such nonlocality is associated with the thermal motion of the particles of the medium. Simple examples of these phenomena in plasmas are illustrated in Subsection 2.4.2.

2.4.1 Magneto-optics of extraordinary polarized anharmonic EM fields

Analysis of the propagation of extraordinary polarized nonsinusoidal fields in magnetized media reveals the new time scales t_{c} for such fields. These time scales prove to depend on the direction of propagation of the fields. In spite of the similarity in mathematical treatment of the problems, characterized by eqns (2.73) and (2.75), these problems are considered separately below, owing to the difference in physical phenomena described by these equations.

To outline the characteristics of the propagation of nonsinusoidal electromagnetic fields in a lossless plasma, dictated by the influence of a magnetic field, let us consider here a strong magnetic field when the characteristic timescale of a travelling wave t_c obeys the condition

$$(\omega_H t_c)^2 \gg 1. \tag{2.76}$$

(A) Longitudinal propagation Let us rewrite the general equations (1.45)–(1.46) using the components of the vector potential A ($A_z = 0$, $A_x \neq 0$, $A_y \neq 0$):

$$\left.\begin{aligned} \frac{\partial^2 A_x}{\partial z^2} - \frac{\varepsilon_\infty}{c^2} \cdot \frac{\partial^2 A_x}{\partial t^2} &= -\frac{4\pi}{c^2} \cdot \dot{j}_x, \\ \frac{\partial^2 A_y}{\partial z^2} - \frac{\varepsilon_\infty}{c^2} \cdot \frac{\partial^2 A_y}{\partial t^2} &= -\frac{4\pi}{c^2} \cdot \dot{j}_y. \end{aligned}\right\} \tag{2.77}$$

Considering the system (2.77) for the case of harmonic waves with a frequency ω, one could find the usual dispersive equations describing the propagation of ordinary and extraordinary circularly polarized waves with refractive indices n_1 and n_2 respectively:

$$n_{1,2}^2 = 1 - \frac{\Omega_e^2}{\omega(\omega + \omega_H)} \tag{2.78}$$

Introducing the cutoff frequency for the ordinary waves ω_c,

$$\omega_c = \frac{\Omega_e^2}{\omega_H} \tag{2.79}$$

one can see that the magnetoplasma is transparent in the spectral range $\omega < \omega_c$ only for the extraordinary waves, whereas both polarizations can propagate in the high-frequency range $\omega > \omega_c$.

The nonsinusoidal fields are analysed by means of the same system (2.77). Differentiating this system twice with respect to t and keeping in mind the condition (2.76), we can rewrite this system in dimensionless form, making use of normalized functions φ_1, φ_2 and variables η and τ:

$$\left.\begin{aligned}
&\varphi_1 = \frac{A_x}{A_0}, \quad \varphi_2 = \frac{A_y}{A_0}, \quad \eta = \frac{z}{v_0 t_{\mathrm{m}}}, \quad \tau = \frac{t}{t_{\mathrm{m}}}, \\
&t_{\mathrm{m}} = -\frac{2(1+\Omega_{\mathrm{e}}^2/\omega_H^2]}{\omega_{\mathrm{c}}}, \quad v_0 = \frac{c}{\sqrt{[\varepsilon_\infty(1+\Omega_{\mathrm{e}}^2/\omega_H^2)]}}, \\
&\frac{\partial^2\varphi_1}{\partial\eta^2} - \frac{\partial^2\varphi_1}{\partial\tau^2} = 2\cdot\frac{\partial\varphi_2}{\partial\tau}, \\
&\frac{\partial^2\varphi_2}{\partial\eta^2} - \frac{\partial^2\varphi_2}{\partial\tau^2} = 2\cdot\frac{\partial\varphi_1}{\partial\tau}.
\end{aligned}\right\} \tag{2.80}$$

The system (2.80) can be reduced to one equation determining both functions φ_1 and φ_2 (Shvartsburg, 1994):

$$\frac{\partial^2\varphi_{1,2}}{\partial\eta^2} - \frac{\partial^2\varphi_{1,2}}{\partial\eta^2} = \pm 2\mathrm{i}\cdot\frac{\partial\varphi_{1,2}}{\partial\tau}. \tag{2.81}$$

The solution of eqn (2.81) may be written as

$$\varphi_{1,2} = \mathrm{e}^{\pm\mathrm{i}\tau} f(\tau, \eta). \tag{2.82}$$

Here f(τ, η) is the nonseparable solution of the KG equations (2.26)–(2.27). Finally, using the definitions (2.5), one can find the electric (e_x, e_y) and magnetic (h_x, h_y) components of the field (2.82) in the form (2.36); here

$$\left.\begin{aligned}
(e_{x,y})_q &= \frac{\mathrm{e}^{\pm\mathrm{i}\tau}}{2}\left[\Psi_{q-1}(\tau,\eta) - \Psi_{q+1}(\tau,\eta) \pm 2\mathrm{i}\Psi_q(\tau,\eta)\right], \\
(h_{x,y})_q &= \frac{\mathrm{e}^{\pm\mathrm{i}\tau}}{2}\left[\Psi_{q-1}(\tau,\eta) + \Psi_{q+1}(\tau,\eta)\right].
\end{aligned}\right\} \tag{2.83}$$

Some features of this time-domain solution should be outlined:

1. Similar to a circularly polarized harmonic wave, the amplitudes $|e_x|_q$ and $|e_y|_q$ relevant to both '+' and '−' branches in eqn (2.83) are equal for each harmonic e_q; the magnetic amplitudes $|h_x|_q$ *and* $|h_y|_q$ are also equal.
2. The temporal variations of the polarization state of each qth harmonic at any fixed position $\eta = \eta_0$ are nonmonotonic.
3. The characteristic time scales t_{c} and spatial scales (2.80) are equal for both '+' and '−' branches of the field (2.83).

The condition (2.76) for the propagation of a circularly polarized nonseparable field may be fulfilled for example in the n-InSb plasma ($\varepsilon_\infty = 15$, electron density $N_{\mathrm{e}} = 8\times10^{14}\,\mathrm{cm}^{-1}$, Langmuir frequency $\Omega_{\mathrm{e}} = 4$

$\times 10^{12}\,\mathrm{rad\,s^{-1}}$), $H_0 = 0.7\,\mathrm{T}$. The time scale t_c (eqn 2.80) in this case is about ~ 1 ps.

(B) Transverse propagation The propagation of an EM field in this geometry is known to depend upon its polarization. Thus the magnetic field has no influence on the ordinary polarized wave with the electric component $\boldsymbol{E} \| \boldsymbol{H}_0$. To describe this wave, one should use the results of previous sections, relevant to a nonmagnetized solid plasma. On the other hand, the propagation of extraordinary field in the x-direction across the magnetic field (z-axis) is described by the set of equations

$$\left.\begin{aligned} \frac{\partial E_y}{\partial x} &= -\frac{1}{c}\cdot\frac{\partial H_z}{\partial t}, \\ -\frac{\partial H_z}{\partial x} &= \frac{4\pi}{c} j_y + \frac{\varepsilon_\infty}{c}\frac{\partial E_y}{\partial t}, \\ 4\pi j_x + \varepsilon_\infty &\frac{\partial E_x}{\partial t} = 0. \end{aligned}\right\} \tag{2.84}$$

Eliminating the magnetic component H_z from the first pair of eqn (2.84), we obtain

$$\frac{\partial^2 E_y}{\partial x^2} - \frac{\varepsilon_\infty}{c^2}\cdot\frac{\partial^2 E_y}{\partial t^2} = \frac{4\pi}{c^2}\cdot\frac{\partial j_y}{\partial t}. \tag{2.85}$$

Supposing the magnetic field to be strong ($p = \Omega_e^2 \cdot \omega_H^{-2} \ll 1$), one can use the expansion of eqn (2.73) with respect to a small parameter p:

$$j_x = \frac{\varepsilon_\infty}{4\pi}\left(\omega_c E_y + p\frac{\partial E_x}{\partial t}\right), \quad j_y = \frac{\varepsilon_\infty}{4\pi}\left(-\omega_c E_x + p\frac{\partial E_y}{\partial t}\right). \tag{2.86}$$

Here $\omega_c = \Omega_e^2 \cdot \omega_H^{-1}$ is the cutoff frequency of the extraordinary sinusoidal wave. Let us differentiate eqn (2.85) twice with respect to t, assuming the characteristic time scale of the travelling field $t_\perp$ to be limited by analogy with eqn (2.76): $(\omega_H t_\perp)^2 \gg 1$. Use of eqn (2.86) yields the dimensionless KG equation (2.12) for the normalized vector potential $f = A_y A_0^{-1}$. Here

$$x = v \cdot t_\perp \cdot \eta, \quad \tau = t/t_\perp, \quad v = c \cdot \varepsilon^{-\frac{1}{2}}, \quad t_\perp = \omega_c^{-1}. \tag{2.87}$$

Unlike the transversely polarized extraordinary EM wave travelling along the magnetic field, the polarization structure of the nonseparable field discussed contains both transversely and longitudinally polarized electric components. The transverse electric and magnetic components of this field, E_y and H_z, are described by the formulae (2.36)–(2.37). The longitudinal electric component E_x is determined by substitution of the current j_x from eqn (2.86) in the last equation of the system (2.84):

$$E_x = \frac{\omega_\perp}{c} A_y. \tag{2.88}$$

In contrast to transverse E_y component, expressed via the derivatives of vector potential A_y, the longitudinal component E_x is expressed via this potential itself. Thus the parity of functions Ψ_q forming the E_y and E_x components differs. Substitution of the value $t_\perp$ (eqn 2.87) in the condition $(\omega_H t_\perp)^2 \gg 1$ leads to the unequality $\omega_H^4 \cdot \Omega_e^{-4} \gg 1$, which is fulfilled by the above-mentioned condition $p = \Omega_e^2 \cdot \omega_H^{-2} \ll 1$.

One should emphasize the anisotropy of the characteristic time scales of nonseparable fields releting to longitudinal propagation (extraordinary polarization, $t_m \gg t_c$ (eqn 2.80) and transverse propagation (ordinary polarization, $t_0 = \Omega_e^{-1}$); the value of the ratio $t_m t_0^{-1}$ is ~4–6. This anisotropy provides the variety of angular- and polarization-dependent time scales for nonseparable fields in magnetoplasmas.

(C) Time-domain solutions of the Landau–Lifshitz equation Consideration of electromagnetic wave phenomena in nonconducting magnetized media, such as ferrites or magnetodielectrics, is usually based on the Landau–Lifshitz equation describing the temporal evolution of the magnetic moment $\boldsymbol{M}$ in the magnetic field $\boldsymbol{H}$:

$$\frac{\partial \boldsymbol{M}}{\partial t} = \gamma [\boldsymbol{H} \cdot \boldsymbol{M}]. \tag{2.89}$$

Here γ is the gyromagnetic ratio $\gamma = eg(2mc)^{-1}$, where g is the Lande factor. The equation governing the magnetic and electric components of an alternating field is obtained by substitution of eqn (2.75) in the Maxwell equation (1 .46):

$$\operatorname{rot} \boldsymbol{H} = \frac{\varepsilon_\infty}{c} \frac{\partial \boldsymbol{E}}{\partial t}. \tag{2.90}$$

Let us consider the alternating perturbations of vectors $\boldsymbol{M}$ and $\boldsymbol{H}$,

$$\boldsymbol{M} = \boldsymbol{M}_0 + \boldsymbol{m}, \quad \boldsymbol{H} = \boldsymbol{H}_0 + \boldsymbol{h}, \quad |\boldsymbol{m}| \ll |\boldsymbol{M}_0|, \quad |\boldsymbol{h}| \ll |\boldsymbol{H}_0|,$$

corresponding to the longitudinal and transverse propagation of the EM field with respect to the direction $\boldsymbol{H}_0 \| \boldsymbol{M}_0$ (z-axis).

Longitudinal propagation of a circularly polarized field, similar to the above-mentioned wave, eqn (2.83), can be analysed readily. However, It is more interesting to consider the transverse propagation of the extraordinary TE polarized field, travelling in the x-direction. This field is characterized by an electric component e_x, magnetic components h_y and h_z, and an alternating magnetic moment m_x. One pair of equations, relating these functions, can be derived from eqns (1.45) and (2.89):

$$\left.\begin{aligned}\frac{\partial e_z}{\partial x}-\frac{1}{c}\cdot\frac{\partial h_y}{\partial t}+\frac{\omega_{\mathrm{m}}}{c}\cdot h_x-\frac{4\pi\omega_H}{c}\cdot m_x=0,\\ \frac{\partial h_x}{\partial t}+\omega_{\mathrm{m}}h_y=0.\end{aligned}\right\}\tag{2.91}$$

Here the frequency ω_{m} is

$$\omega_{\mathrm{m}}=4\pi\gamma M_0.\tag{2.92}$$

The other pair of equations follows from eqns (1.47) and (2.90):

$$\left.\begin{aligned}\frac{\partial h_x}{\partial x}+4\pi\frac{\partial m_x}{\partial x}=0,\\ \frac{\partial h_y}{\partial x}-\frac{\varepsilon_\infty}{c}\frac{\partial e_z}{\partial t}=0.\end{aligned}\right\}\tag{2.93}$$

Since the complex system (2.91), (2.93) contains, unlike the above-mentioned problems, four unknown functions, it is expedient to vary the standard scheme of obtaining nonseparable solutions of this system. Let us introduce the auxiliary function f and the normalization constant A_0 so that

$$e_z=A_0\cdot\frac{\partial f}{\partial x},\quad h_y=\frac{\varepsilon_\infty}{c}\cdot A_0\cdot\frac{\partial f}{\partial t}.\tag{2.94}$$

Through this substitution the second of eqns (2.93) becomes an identity, while the system (2.91) can be transformed to a dimensionless form of the KG equation (2.12); here

$$\eta=\frac{x}{vt_{\mathrm{c}}},\quad \tau=\frac{t}{t_{\mathrm{c}}},\quad v=c\cdot\varepsilon_\infty^{-1/2},\quad t_{\mathrm{c}}=[\omega_{\mathrm{m}}(\omega_{\mathrm{m}}+\omega_H)]^{-\frac{1}{2}}.\tag{2.95}$$

The function f and the components of the nonseparable field may be written as

$$\left.\begin{aligned}f=\sum_{q=1}^{\infty}a_q\cdot\Psi_q(\tau,\eta),\quad e_z=\frac{A_0}{vt_{\mathrm{c}}}\sum_{q=1}^{\infty}a_q\frac{\partial\Psi_q}{\partial\eta},\\ h_y=\frac{\sqrt{\varepsilon_\infty}A_0}{vt_{\mathrm{c}}}\sum_{q=1}^{\infty}a_q\frac{\partial\Psi_q}{\partial\tau},\quad h_x=-\frac{\omega_{\mathrm{m}}\sqrt{\varepsilon_\infty}A_0}{v}\sum_{q=1}^{\infty}a_q\Psi_q,\\ m_x=-\frac{1}{4\pi}h_x.\end{aligned}\right\}\tag{2.96}$$

The dispersive equation for the harmonic wave described by the system (2.91)–(2.93) is

$$k^2v^2=\omega^2-\omega_{\mathrm{m}}(\omega_{\mathrm{m}}+\omega_H).\tag{2.97}$$

Hence the time scale of the nonseparable TE polarized field t_c (eqn 2.95) is determined by the cutoff frequency of harmonic wave (eqn 2.97). Considering the saturating fields typical for ferrite and ferrite–garnet devices ($\varepsilon_\infty = 9$, $g = 2$), $H_0 = 0.4$–0.5 T, $M_0 = 150$–180 Gs, one can find from eqn (2.95) the time scale $t_c = 20$–25 ps; the relevant spatial scale of the nonseparable fields is $x_c = vt_c = 0.2$–0.25 cm. The parameters t_c and x_c prove to be anisotropic.

The dependence of the spatial–temporal scales of nonstationary EM fields travelling in magnetized media upon the external magnetic field and geometry of propagation demonstrates the possibilities of controlled tuning of these scales in the subpicosecond and picosecond ranges. The depth of effective penetration of such nonseparable field harmonics (Fig. 2.4) may be discerned at several times through the angular and polarization dependences of the spatial scale x_c. These possibilities provide for the development of time-domain magneto-optics and radio-optics of nonsinusoidal fields.

2.4.2 Anharmonic waves in media with spatial dispersion

To discover the influence of nonlocality on the spatial–temporal structure of anharmonic fields, let us reconsider the motion of plasma electrons, the finite temperature of the electron gas T_e being taken into account. The equation governing the electron current $\boldsymbol{J}$ in nonmagnetized plasma may be written, unlike eqn (2.4), in the form

$$\frac{\partial \boldsymbol{J}}{\partial t} = \frac{\omega_p^2 \boldsymbol{E}}{4\pi} - \frac{3ke}{mN_e} \nabla (T_e N_e). \tag{2.98}$$

The last term in eqn (2.98) describes the pressure of the electron gas characterized by temperature T_e and density N_e; k is the Boltzmann constant. The perturbation of density N_1 produced by an electric field $\boldsymbol{E}$ obeys the Poisson equation

$$\operatorname{div} \boldsymbol{E} = 4\pi e N_1. \tag{2.99}$$

Considering the longitudinal electric field E_z and making use of eqns (2.98) and (2.99), we obtain

$$\frac{\partial j_z}{\partial t} = \frac{\omega_p^2 E_z}{4\pi} - \frac{3v_T^2}{4\pi} \cdot \frac{\partial^2 E_z}{\partial z^2}. \tag{2.100}$$

Here $v_T = \sqrt{kT_e m^{-1}}$ is the thermal velocity of the electron carriers. On the other hand, the electric current $\boldsymbol{J}$ induced by the curl-less electric field E_z(rot $\boldsymbol{H} = 0$) may be expressed by means of eqn (1.46) as

$$j_z = -\frac{\varepsilon_\infty}{4\pi} \cdot \frac{\partial E_z}{\partial t}. \tag{2.101}$$

Substitution of eqn (2.101) in eqn (2.100) yields the equation describing the plasma waves:

$$\frac{\partial^2 E_z}{\partial z^2} - \frac{\varepsilon_\infty}{3v_T^2} \cdot \frac{\partial^2 E_z}{\partial t^2} = \frac{\omega_p^2}{3v_T^2} \cdot E_z. \tag{2.102}$$

The harmonic plasma waves $E_z \approx \exp[i(kz - \omega t)]$ are known to obey the dispersive equation, following from eqn (2.102),

$$\omega = \Omega_e \left[1 + \frac{3}{2} (kr_D)^2 \right]. \tag{2.103}$$

Here $r_D = v_T \omega_p^{-1}$ is the Debye radius, determining the nonlocality scale in the plasma; this scale provides the finite length of resonant plasma oscillations.

Introducing the normalized variables

$$\eta = \frac{z}{r_D 3^{1/2}}, \quad \tau = \Omega_e t, \quad E_z = E_0 f(\tau, \eta), \tag{2.104}$$

one can rewrite eqn (2.102) in a form of the KG equation (2.12) for the function $f(\tau, \eta)$. The nonsinusoidal plasma wave governed by this equation may be written as

$$f(\tau, \eta) = \sum_{q=1}^{\infty} a_q \Psi_q. \tag{2.105}$$

Unlike the harmonic plasma wave, the propagation of the field (2.105) is accompanied by the reduction of its maximum and complex oscillations of its velocity (Fig. 2.4). These oscillations violate the conditions of resonant Landau damping of plasma waves, since the synchronous motion of electrons and field maxima becomes impeded. The effective depth of penetration of such fields into a collisionless plasma is determined by dispersive processes.

The solution (2.105) is valid for isotropic plasma an well as for the field travelling along the magnetic field H_0 (z-axis) in a magnetized plasma. A similar nonseparable solution can be obtained for a field E_x propagating in the x-direction across the z-axis. The equation governing the current component j_x in a magnetoplasma may be written

$$\frac{\partial j_x}{\partial t} = \frac{\omega_p^2 E_x}{4\pi} + \omega_H j_y - \frac{3ke}{mN_e} \frac{\partial}{\partial x} (T_e N_e). \tag{2.106}$$

Combining the second equation from the system (2.73) ($E_y = 0$) with eqn (2.84), we find

$$j_y = \frac{\omega_H}{4\pi} E_x. \tag{2.107}$$

In spite of the absence of electric field component E_y, a nonzero value of electric current component j_y (eqn 2.107) is provided by the gyrotropy effect. Substitution of eqn (2.107) in eqn (2.106) and use of the normalized variables

$$\eta = \frac{x}{r_D 3^{1/2}}, \quad \tau = \sqrt{(\omega_H^2 + \Omega_e^2)}t \quad E_x = E_0 f(\tau, \eta) \qquad (2.108)$$

again yields the KG equation (2.12) with a new time scale (2.108); here the solution (2.105) remains valid, the ratio between the frequencies Ω_e and ω_H being arbitrary. The frequency $\sqrt{(\omega_H^2 + \Omega_e^2)}$ is known as the upper hybrid resonant frequency of a magnetoplasma.

Noting some analogies above between the sinusoidal and nonsinusoidal fields in transparent magnetized media, described by the same KG equation, one would emphasize the important distinctions between these fields:

1. no cutoff frequency is defined for nonseparable fields in transparent dielectrics.
2. unlike sinusoidal waves travelling in opposite directions along the magnetic field in some magnetized dielectrics, the nonseparable fields in these media do not possess nonreciprocal properties.
3. Since the aforesaid anharmonic fields have no phase velocity, such fields travelling in magnetized media (eqn 2.83) do not possess Faraday rotation of the polarization state.

2.5 Exact time-domain solutions of the telegraph equation

The telegraph equation and its frequency-domain solutions describing electromagnetic phenomena in conducting media laid the foundations of traditional analysis of electrical power-carrying and communication systems, two-wire Lecher lines and coaxial waveguides. A similar approach, developed in the electrodynamics of bulk absorbing media, results in the use of attenuation decrements of sinusoidal waves, associated with the imaginary parts of the frequencies and wavenumbers of such waves. The classical picture of these phenomena was investigated several decades ago. On the other hand, the recent successes in the development of microstrip transmission lines for ultrashort picosecond and subpicosecond transients (Qian and Yamashita 1991) have stimulated the reconsideration of these nonstationary phenomena in the time domain. Some trends in the interaction of EM transients with lossy media can also be visualized by the time-domain representation (Papazoglou 1975). However, such representations were obtained by means of inverse Fourier transforms of solutions found previously in the frequency domain.

In contrast, direct, exact time-domain solutions of the telegraph equation are advanced in this section. These results can be considered as a generalization of nonseparable fields in lossless media, discussed above. Neither the real parts of frequencies and wavenumbers nor their imaginary parts are used in the analysis of such nonsinusoidal fields; thus the complicated distortions of these fields in lossy systems are not described in the time domain by the usual attenuation decrement or absorption length. Similarly to transparent media, the rates of evolution of the electric and magnetic components of an EM field are also different in absorbing systems. Unlike in transparent media, the spatial–temporal structures of the harmonics of EM fields in conducting media are distinguished by the absence of zero-crossing points. To reveal these special features, the propagation of nonseparable fields in a bulk conducting material is considered first (Subsection 2.5.1). The same approach is developed in Subsection 2.5.2 for the analysis of nonsinusoidal fields in distributed lossy systems and transmission lines.

2.5.1 Anharmonic electrodynamics of conducting media

Let us consider the spatial–temporal structure of electromagnetic fields in media obeying Ohm's law,

$$J = \sigma \cdot E, \tag{2.109}$$

where σ is the conductivity of the medium. Examining for example a linearly polarized field with components E_x and H_y travelling in the z-direction, one can derive from eqns (1.45)–(1.46) and eqn (2.109) the equation governing e.g. the E_x component:

$$\frac{\partial^2 E_x}{\partial z^2} - \frac{\varepsilon_\infty}{c^2} \cdot \frac{\partial^2 E_x}{\partial t^2} = \frac{4\pi\sigma}{c^2} \cdot \frac{\partial E_x}{\partial t}. \tag{2.110}$$

In the model of harmonic waves, eqn (2.110) is known to describe the damping oscillations. The attenuation of the sinusoidal oscillations of both electric and magnetic components is characterized in this model by the same e-folding length L_a. Introducing the time scale of the medium T_σ,

$$T_\sigma = \frac{\varepsilon_\infty}{2\pi\sigma}, \tag{2.111}$$

one can find the attenuation length L_a for a monochromatic wave with frequency ω:

$$L_a = \frac{v}{\omega} \cdot \sqrt{\left(\frac{2}{-1 + \sqrt{[1 + 4(\omega T)^{-2}]}}\right)}, \tag{2.112}$$

where $v = c \cdot \varepsilon_\infty^{-\frac{1}{2}}$. If the frequency ω is high enough $[(\omega T)^2 \gg 4]$, the distance L_a becomes independent of the frequency; this case relates to the shortest attenuation length L_σ (eqn 2.113). The length L_a increases if the frequencies are low $[\omega T)^2 \ll 4]$.

$$L_\sigma|_{(\omega T)^2 \gg 4} = v \cdot T, \quad L_a|_{(\omega T)^2 \ll 4} = \frac{L_\sigma}{\sqrt{\omega T}}. \tag{2.113}$$

To find the nonseparable solutions of eqn (2.110) it is worth while to use the vector potential A (eqn 2.5) ($A_x \neq 0$, $A_y = A_z = 0$) and the normalized variables $\eta = zL_\sigma^{-1}$, $\tau = tT_\sigma^{-1}$, $f(\tau, \eta) = A_x A_0^{-1}$; here the spatial and temporal scales L_σ and T_σ are defined by eqns (2.113) and (2.111) respectively. Use of these variables reduces eqn (2.110) to the dimensionless telegraph equation:

$$\frac{\partial^2 f}{\partial \eta^2} - \frac{\partial^2 f}{\partial \tau^2} = 2 \cdot \frac{\partial f}{\partial \tau}. \tag{2.114}$$

The solution of eqn (2.114) may be written as

$$f = \mathrm{e}^{-\tau} U(\tau, \eta). \tag{2.115}$$

The function $U(\tau, \eta)$ obeys the so-called 'Klein–Gordon equation with imaginary mass',

$$\frac{\partial^2 U}{\partial \eta^2} - \frac{\partial^2 U}{\partial \tau^2} = -U. \tag{2.116}$$

This equation can be transformed to the 'unual' KG equation (2.12) by the substitutions $\eta_1 = \mathrm{i}\eta$, $\tau_1 = \mathrm{i}\tau$. Generalizing the analysis (2.19)–(2.25) for the imaginary variables η_1 and τ_1, one can find that the solution at the boundary plane $\eta = 0$ may be represented, by analogy with eqn (2.25), as a sum of modified Bessel functions $I_q(u)$ with integer positive indices q. Finally, making use of eqn (2.116), we obtain the vector potential f (eqn 2.115) in the form

$$\left.\begin{aligned} f &= \mathrm{e}^{-\tau} \sum_{q=1}^{\infty} a_q \theta_q(\tau, \eta), \\ \theta_q(\tau, \eta) &= \left(\frac{\tau - \eta}{\tau + \eta}\right)^{q/2} I_q(u), \quad u = \sqrt{\tau^2 - \eta^2}. \end{aligned}\right\} \tag{2.117}$$

This solution is valid in the area $\tau \geqslant \eta$. The function θ_0,

$$\theta_0 = I_0(\sqrt{\tau^2 - \eta^2}) \tag{2.118}$$

is the Green's function for eqn (2.116).

The analysis of these solutions is similar to those in Section 2.2. Thus, use of the formula (A.2.3) yields the following expressions for the nonseparable components of an EM field E_x and H_y in a conducting medium (Shvartsburg 1995):

$$E_x = -\frac{A_0}{cT_\sigma} \cdot \sum_{q=1}^{\infty} a_q e_q, \quad H_y = -\frac{A_0}{cT_\sigma} \cdot \sum_{q=1}^{\infty} a_q h_q, \tag{2.119}$$

$$e_q = \frac{e^{-\tau}}{2} [\theta_{q-1}(u) + \theta_{q+1}(u) - 2\theta_q(u)], \tag{2.120}$$

$$h_q = \frac{e^{-\tau}}{2} [\theta_{q-1}(u) - \theta_{q+1}(u)]. \tag{2.121}$$

One can be convinced by means of asymptotic expansion of the modified Bessel functions (A.2.4) that the field distributions (2.120), (2.121) tend to zero at the wings of these distributions ($u \gg 1$).

Examples of the spatial–temporal evolution of the harmonics of the nonsinusoidal field (2.120)–(2.121) are shown in Fig. 2.8. The spatial scale of these graphs relates to the shortest attenuation length L_σ (eqn 2.113). The following features of the time-domain description of nonseparable fields in the conducting media should be outlined:

1. Unlike harmonic waves, the distortions of the fields (2.120)–(2.121), being heterogeneous, are not characterized by only one parameter, e.g. by an attenuation length (eqn 2.112). The envelope deformations at the wings of the distributions of the electric and magnetic components are smaller than in their central parts, since the peripheral reduction of these fields is compensated partly by spreading of the maxima.
2. The domains of localization of the electric components are two to three times narrower than those of the magnetic components.
3. The amplitudes of higher e_q and h_q harmonics ($q = 2, 3, 4, \ldots$) at fixed space–time points prove to be much smaller than the corresponding values for the first harmonics ($q = 1$).
4. The temporal envelopes of harmonics e_q and h_q possess no oscillations in the areas of effective field localization.
5. Owing to heterogeneous distortions of harmonics e_q and h_q in the course of propagation, the moments when their peak amplitudes reach any fixed position η within the medium are spaced in time.

Since the telegraph equation is used in many fields of physics, these exact time-domain solutions will be shown below to have a series of applications both in electrodynamics and in cross-disciplinary physics.

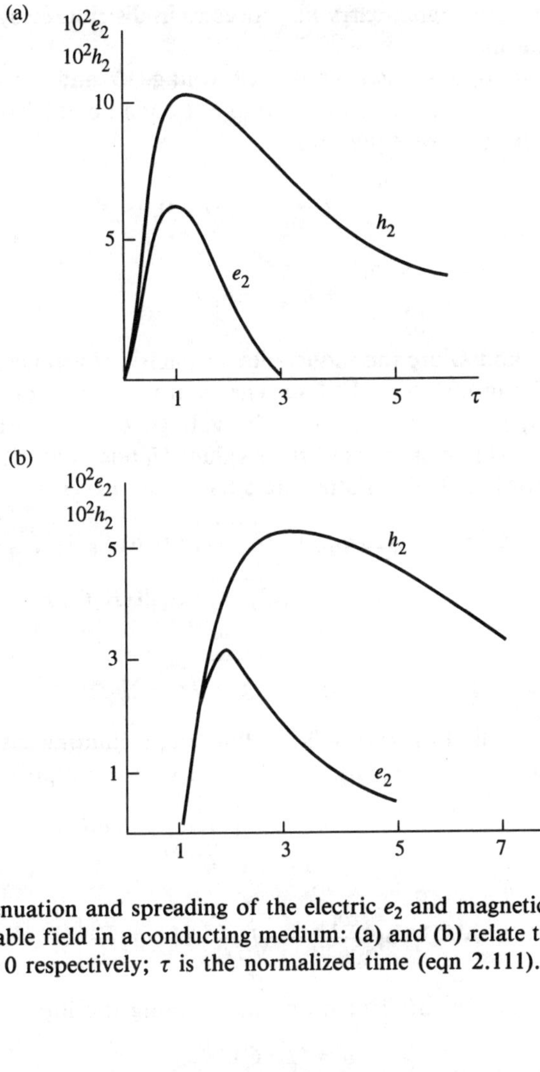

Fig. 2.8 Attenuation and spreading of the electric e_2 and magnetic h_2 harmonics of a nonseparable field in a conducting medium: (a) and (b) relate to the positions $\eta = 0$ and $\eta = 0$ respectively; τ is the normalized time (eqn 2.111).

2.5.2 Nonstationary fields in transmission lines

The telegraph equation (2.114) is used traditionally for analysis of processes in transmission lines and guiding electromagnetic systems characterized by distributed resistance, inductance, and capacity. Therefore the nonseparable solutions of this equation, eqns (2.120)–(2.121), obtained for conducting half-space may be generalized properly for the development of nonstationary electrodynamics of transmission lines. The time-domain reconsideration of this classical problem can provide a new interpretation of the multitude of energy-carrying and signal-carrying processes in distributed systems similar to transmission lines.

The spatial–temporal distributions of voltage U and electric current I in a standard model of a transmission line (Landau and Lifshitz 1984) are governed by the pair of equations

$$\frac{\partial I}{\partial z} + C\frac{\partial U}{\partial t} + GU = 0, \tag{2.122}$$

$$\frac{\partial U}{\partial z} + L\frac{\partial I}{\partial t} + RI = 0. \tag{2.123}$$

Here L, C, R, and G are the inductance, capacity, resistance, and leakage parameter per unit length of the line. The traditional solution of the system (2.122)–(2.123) is known to express the voltage U_2 and current I_2 at the loaded end of this line in terms of their values U_1 and I_1 at the input end by means of a stationary train of attenuated harmonic waves with frequency ω:

$$U_2 = U_1 \mathrm{ch}(R_0 G_0) - \sqrt{R_0/G_0} \cdot I_1 \mathrm{sh}(R_0 G_0), \tag{2.124}$$

$$I_2 = I_1 \mathrm{ch}(R_0 G_0) - \sqrt{G_0/R_0} \cdot U_1 \mathrm{sh}(R_0 G_0). \tag{2.125}$$

Here

$$R_0 = (R + \mathrm{i}\omega L)l, \quad G_0 = (G + \mathrm{i}\omega C)l, \tag{2.126}$$

where l is the length of the transmission line. In the limiting case of a lossless line ($R \to 0$, $G \to 0$) the solution (2.124)–(2.125) is simplified:

$$U_2 = U_1 \cos\left(\frac{\omega l}{v}\right) - \mathrm{i}\sqrt{\left(\frac{L}{C}\right)}\, I_1 \sin\left(\frac{\omega l}{v}\right), \tag{2.127}$$

$$I_2 = I_1 \cos\left(\frac{\omega l}{v}\right) - \mathrm{i}\sqrt{\left(\frac{L}{C}\right)}\, U_1 \sin\left(\frac{\omega l}{v}\right). \tag{2.128}$$

Here v is the velocity of field propagation along the line,

$$v = (L \cdot C)^{-\frac{1}{2}}. \tag{2.129}$$

In contrast to this stationary solution, let us consider the nonstationary fields described by the same system (2.122)–(2.123). Defining the characteristic time scales

$$t_1 = L/R, \quad t_2 = C/G, \quad T_0 = 2t_1 t_2/|t_1 - t_2|, \tag{2.130}$$

and making use of the normalized voltage u, current i and variables η and τ,

$$u = \frac{UA_0}{R}, \quad i = vTA_0 I, \quad \eta = \frac{z}{vT_0}, \quad \tau = \frac{t}{T_0}, \tag{2.131}$$

one can rewrite the system (2.122)–(2.123) in the dimensionless form

$$\frac{t_1 t_2}{T_0^2} \cdot \frac{\partial i}{\partial \eta} + \frac{t_2}{T_0} \cdot \frac{\partial u}{\partial \tau} + u = 0, \tag{2.132}$$

$$\frac{\partial u}{\partial \eta} + \frac{t_1}{T_0} \cdot \frac{\partial_1}{\partial \tau} + i = 0. \tag{2.133}$$

The velocity v in eqn (2.131) is determined by eqn (2.129).

Let us express the normalized voltage u and current i via an auxiliary function F so that

$$i = \frac{\partial F}{\partial \eta}, \quad u = -\frac{t_1}{T_0} \frac{\partial F}{\partial \tau} - F. \tag{2.134}$$

Substitution of eqns (2.134) in eqn (2.132) transforms the latter to an identity, while eqn (2.133) yields the equation governing the function F*:

$$\frac{\partial^2 F}{\partial \eta^2} - \frac{\partial^2 F}{\partial \tau^2} - \frac{T_0(t_1 + t_2)}{t_1 t_2} \frac{\partial F}{\partial \tau} = \frac{T_0^2}{t_1 t_2} F. \tag{2.135}$$

Representing F in the form

$$F = \mathrm{e}^{-\alpha\tau} f, \quad \alpha = \frac{T_0(t_1 + t_2)}{2 \cdot t_1 t_2}, \tag{2.136}$$

we can finally reduce eqn (2.135) to the Klein–Gordon equation with imaginary mass, eqn (2.116). Substitution of the solution of this equation, expressed by the functions $\theta_q(\tau, \eta)$ (eqn 2.117), in eqns (2.136) and (2.134) results in explicit formulae describing the nonsinusoidal electric current i and voltage u in a transmission line:

$$i = -\sum_{q=1}^{\infty} a_q i_q, \quad u = -\sum_{q=1}^{\infty} a_q u_q, \tag{2.137}$$

$$i_q = \frac{1}{2} \mathrm{e}^{-\alpha\tau} [\theta_{q-1}(u) - \theta_{q+1}(u)], \tag{2.138}$$

$$u_{q+} = u_q|_{t_2 > t_1} = \frac{1}{4} \left| \frac{t_1}{t_2} - 1 \right| \mathrm{e}^{-\alpha\tau} [\theta_{q-1}(u) + \theta_{q+1}(u) + 2\theta_q(u)], \tag{2.139}$$

---

*Using, instead of eqn (2.134), another substitution,

$$u = \frac{t_1}{T_0} \frac{\partial F}{\partial \eta}, \quad i = -\frac{\partial F}{\partial \tau} - \frac{T_0}{T_2} F,$$

one can reduce eqn (2.133) to an identity, while eqn (2.132) again becomes eqn (2.135).

$$u_{q-} = u_q\big|_{t_2<t_1} = \frac{1}{4}\left|\frac{t_1}{t_2} - 1\right| e^{-\alpha\tau}[\theta_{q-1}(u) + \theta_{q+1}(u) - 2\theta_q(u)]. \quad (2.140)$$

The system (2.122)–(2.123) is characterized by two time scales t_1 and t_2 (eqn 2.130) and, subject to their ratio ($t_2 > t_1$ or $t_2 < t_1$), the nonseparable harmonic of voltage u_q is determined either by eqn (2.139) or by eqn (2.140). The case $t_1 = t_2(T \to \infty)$ relates to the shape-preserving propagation of an arbitrary initial field envelope, the e-folding distance being equal to $z = vt_1$.

The processes of evolution of nonseparable EM fields in conducting bulk media and transmission lines possess many similar trends. Thus there are some analogies between the behaviour of the magnetic component of a field h_q (eqn 2.121) and the electric current i_q (eqn 2.138), while the electric component e_q of this field (eqn 2.120) proves to be similar to the voltage u_{q-} (eqn 2.140).

The nonstationary harmonics of current and voltage do not possess any zero-crossing points. However, unlike in a bulk medium, the presence of two independent time scales t_1 and t_2 (eqn 2.130) provides the formation of either $(u_q)_+$ or $(u_q)_-$ amplitudes of voltage, eqns (2.139)–(2.140). These analogies offer the possibility of flexible modelling of attenuating nonseparable fields in transmission lines (Section 3.6.2).

2.6 Guiding systems for nonseparable fields

Anharmonic EM fields in transparent and absorbing media, described above by the nonseparable solutions (2.27) and (2.117), relate to plane waves in one-dimensional geometry. However, the direct generalization of these solutions for 2D and 3D problems in continuous media proves to be difficult, although some examples of nonseparable 3D fields in free space were found several decades ago (Bateman 1955). Therefare 2D and 3D fields may be represented in dispersive media by means of two types of 'mixed' solutions:

(1) the product of coordinate-dependent functions in the nonseparable solution of the one-dimensional Klein–Gordon or telegraph equation;
(2) the product of sinusoidal time-dependent functions in the stationary nonseparable solution of the Helmholtz equation.

The first approach may be of interest for analysis of nonstationary anharmonic fields in guiding systems (Subsection 2.6.1). These fields are now attracting attention as a result of successes in the development of sources for terahertz optics and radio-optics of picosecond waveforms (Fattinger and Grischkowsky 1989). Such sources usually contain coplanar microstrip

(a)

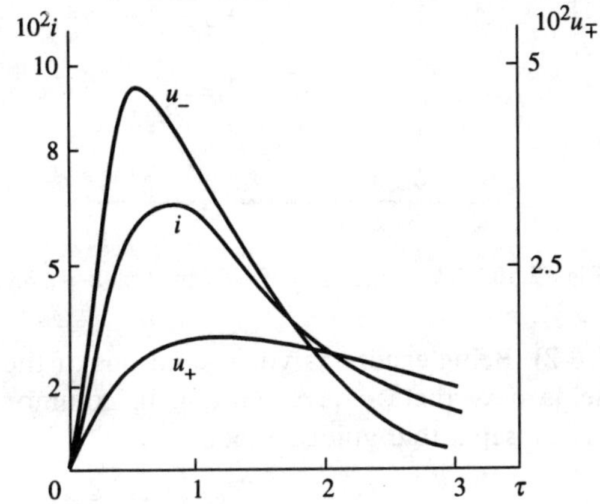

(b)

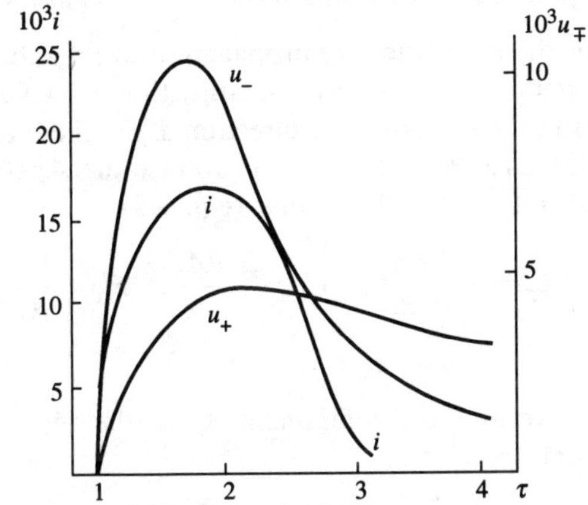

Fig. 2.9 Attenuation and spreading of nonsinusoidal harmonics of current i_2 and voltage $(u_2)_{\pm}$ in a transmission line at the position (a) $\eta = 0$ and (b) $\eta = 1$. The graphs u_+ and u_- relate to values of the ratio of time scales $t_1 t_2^{-1} = 0.2$ and $t_1 t_2^{-1} = 5$ respectively (eqn 2.130), τ is the normalized time.

lines or high-T_c superconducting transmission lines bearing the superconducting films deposited on special substrates (Nuss and Goossen 1989). These devices extend radio and microwave techniques into the terahertz region. The distinction between such short electrical and optical pulses is vanishing, and the reconsideration of nonstationary regimes in perfectly conducting striplines becomes worth while.

Approach (2) provides the description of unusual spatially and temporally modulated structures of localized beams of travelling harmonic waves

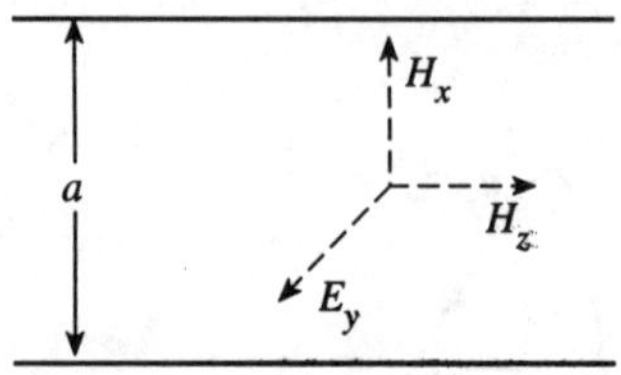

Fig. 2.10 TE_{01} polarized fields in a slab waveguide.

(Subsection 2.6.2). Being exact analytical solutions of the Helmholtz equation, these fields nevertheless have nothing in common with traditional representations of separable guided modes.

2.6.1 Radio-optics of anharmonic modes in waveguides

Let us consider the nonstationary propagation of an anharmonic EM field in a simple model of a guiding system, consisting of two coplanar perfectly conducting plates. Designating the direction along these plates as z and the normal direction as x (Fig. 2.10), we introduce the A_y component of vector potential A ($A_x = A_z = 0$) so that (eqn 2.5)

$$\left.\begin{aligned} E_y = -\frac{1}{c}\frac{\partial A_y}{\partial t}, \quad H_x = -\frac{\partial A_y}{\partial z}, \quad H_z = \frac{\partial A_y}{\partial x}, \\ E_x = E_z = H_y = 0. \end{aligned}\right\} \tag{2.141}$$

The equation governing the component A_y can be derived from the Maxwell equations (1.45)–(1.48):

$$\frac{\partial^2 A_y}{\partial z^2} + \frac{\partial^2 A_y}{\partial x^2} - \frac{1}{c^2}\frac{\partial^2 A_y}{\partial t^2} = 0. \tag{2.142}$$

The standard boundary conditions at the plates $x = \pm b/2$,

$$E_y|_{x=\pm b/2} = 0, \tag{2.143}$$

permit one to search for the solution of eqn (2.142) in the form of a sum of different eigenmodes:

$$A_y = A_0 \sum_{n=0}^{\infty} a_n f_n(z, t) \cos(k_{\perp n} x), \tag{2.144}$$

$$A_{\perp n} = \pi(1 + 2n)b^{-1}. \tag{2.145}$$

The dimensionless functions $f_n(z, t)$ describe the nth mode of the trapped field; each of these modes is determined by the KG equation arising from substitution of eqn (2.144) in eqn (2.142). In the case of a monochromatic wave train, $f_n = \exp[i(\beta_n z - \omega t)]$, one would obtain from eqn (2.142)

$$\beta_n^2 c^2 = \omega^2 - c^2 k_{\perp n}^2. \tag{2.146}$$

Here β_n is the propagation constant of the nth mode. Waveguide equation (2.146) coincides with plasma dispersive equation (2.9) through the replacement $\Omega_e^2 \to c^2 k_{\perp n}^2$, $v^2 k^2 \to \beta_n^2 c^2$. Therefore, instead of one KG equation (2.12) for a plasma-like medium, a multitude of nonseparable equations for anharmonic modes in a waveguide can be written. Introducing the normalized variables

$$\eta_n = z k_{\perp n}, \quad \tau_n = k_{\perp n} ct, \tag{2.147}$$

we obtain from eqn (2.142) the KG equation for the nth mode:

$$\frac{\partial^2 f_n}{\partial \eta_n^2} - \frac{\partial^2 f_n}{\partial \tau_n^2} = f_n. \tag{2.148}$$

The solution of eqn (2.148) may be written, by analogy with eqn (2.26), as a sum of harmonics $\Psi_q(\tau_n, \eta_n)$. Combining eqns (2.26)–(2.27) with eqn (2.144), we present the vector potential A_y as a double sum with respect to mode indices n and harmonic numbers q:

$$A_y = A_0 \sum_{n=0}^{\infty} a_{nq} \cos(k_{\perp n} x) \sum_{q=1}^{\infty} \Psi_q(\tau_n, \eta_n). \tag{2.149}$$

Substitution of eqn (2.149) in eqn (2.141) yields explicit time-domain expressions for the E_y, H_x, and H_z components of the travelling nonsinusoidal modes, e.g.

$$E_{yn} = -A_0 k_{\perp n} \cos(k_{\perp n} x) \sum_{q=1}^{\infty} a_{nq} e_q(\tau_n, \eta_n), \tag{2.150}$$

$$e_q(\tau_n, \eta_n) = \frac{1}{2}[\Psi_{q-1}(\tau_n, \eta_n) - \Psi_{q+1}(\tau_n, \eta_n)]. \tag{2.151}$$

Similarly to the harmonic TE_{01} mode in a coplanar waveguide, the transverse and longitudinal components of the anharmonic mode (2.149), expressed in terms of the functions $\Psi_{q\pm1}$ and Ψ_q respectively, possess different symmetry with respect to the transposition $\tau_n \to -\tau_n$, $\eta_n \to -\eta_n$.

The graphs in Figs 2.1–2.6 illustrating the dispersive phenomena for nonsinusoidal harmonics in a plasma-like medium remain valid for the corresponding components of waveguide modes after the replacement of the normalized variables $\tau_n \to \tau_n$, $\eta_n \to \eta_n$. Hence the higher the number of the nonseparable harmonic q, the faster is its diffusion; this effect common both to a bulk dispersive medium and to a waveguide, is shown in Fig. 2.11 by way of harmonics e_2 and e_4.

However, the real variables characterizing the spatial and temporal scales of diffusion of each mode are different:

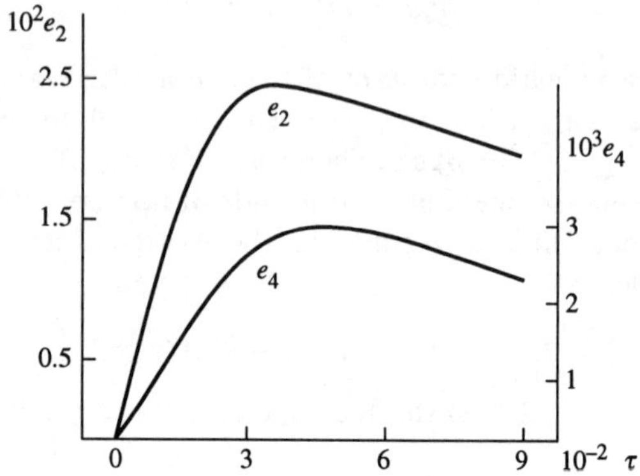

Fig. 2.11 Accelerated diffusion of the envelope of electric component of harmonic e_4 as compared with the envelope e_2 in the depth of a dispersive medium ($\eta = 100$).

$$L_n = \frac{b}{\pi(1+2n)}, \quad t_n = \frac{L_n}{c}. \tag{2.152}$$

This difference provides the special effect of spatial filtration of the higher nonstationary modes, resulting in accelerated diffusion of higher modes. Owing to this filtration, the fundamental mode ($n = 0$) may predominate in the structure of the nonseparable field in the depth of a waveguide (Fig. 2.12). The dispersive spreading of the trapped field leads to the formation of quasi-harmonic oscillations at the wings of the mode distributions (eqn 2.150); the discrete spectrum of these oscillations consists of the mode eigenfrequencies $\omega_n = ck_{\perp n}$.

By evaluating the scales of nonsinusoidal fields in guiding coplanar systems, one can find that for example in a typical case, $a = 10\,\mu\text{m}$, the dimensionless length of propagation $l = 1$ cm corresponds to the value $\eta = 3 \times 10^4$. This system can provide generation of a fundamental eigenfrequency $\omega_0 = 10^{14}\,\text{rad s}^{-1}$, corresponding to the far IR region, with a free-space wavelength $\lambda_0 = 20\,\mu\text{m}$. The higher eigenfrequencies may be estimated as $\omega_n = n\omega_0$.

The fields in such coplanar transmission lines may be considered as 'mixed' 2D structures, associated with distributions of nonstationary fields in the x- and z-directions. The formal generalization of this approach for 3D structures of guided fields is obvious; a physically meaningful example of such generalization, useful for modelling of nonsinusoidal fields in cryoacoustical circularly waveguides, will be discussed later (Section 3.6).

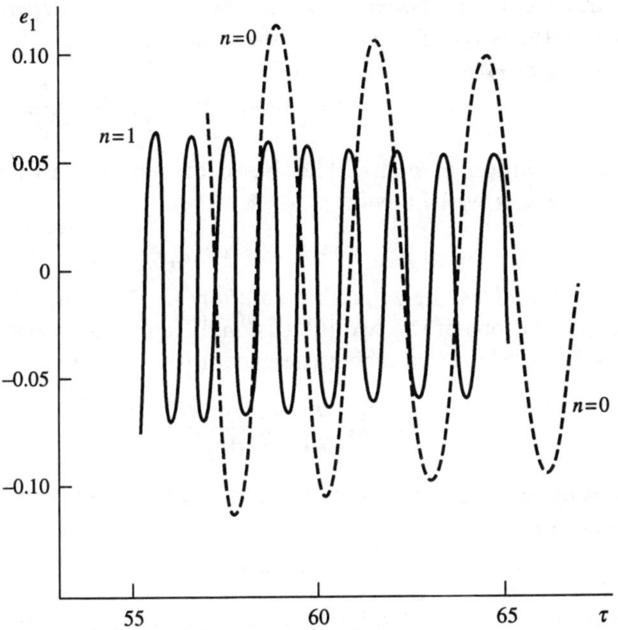

Fig. 2.12 Nonseparable harmonics of electric field e_1 travelling in fundamental ($n = 0$) and higher ($n = 1$) modes in a slab waveguide (Fig. 2.10) at the position $\eta = 50$, the peak amplitudes of both harmonics being equal at the waveguide input ($\eta = 0$).

2.6.2 Nonseparable spatial structures of harmonic modes

This section is devoted to 'mixed' structures of 3D localized fields. Unlike the anharmonic waves discussed above, the spatial–temporal structure of these fields is distinguished by the factor $\exp[i(\beta z - \omega t)]$, typical for sinusoidal waves travelling in the z-direction. To illustrate the special features of these structures it is expedient to examine first the stationary nonseparable problem for a train of sinusoidal waves in a rectangular waveguide. Let us consider the TM_{11} mode with the components E_x, E_y, E_z, H_x, H_y in a waveguide with perfectly conducting walls. The transverse components of the trapped mode are known to be expressed in terms of the longitudinal component E_z. Keeping in mind the travelling TM_{11} wave with frequency ω and expressing the component E_z in the form $E_z = E_0 e(x, y)[i(\beta z - \omega t)]$, we can write the equation governing the coordinate-dependent function $e(x, y)$:

$$\frac{\partial^2 e}{\partial x^2} + \frac{\partial^2 e}{\partial y^2} = k_\perp^2 e. \tag{2.153}$$

Here the transverse component of the wave-vector $k_\perp$ depends on the dimensions of the waveguide cross section a and b:

$$k_\perp^2 = \frac{\omega^2}{c^2} - \beta^2, \quad k_\perp^2 = k_1^2 + k_2^2, \quad k_1 = \frac{\pi}{a}, \quad k_2 = \frac{\pi}{b}. \tag{2.154}$$

The standard solution of eqn (2.153) based on separation of the spatial variables x and y is well known:

$$e_0(x, y) = \cos(k_1 x) \cdot \cos(k_2 y). \tag{2.155}$$

However, as well as the separable solution (2.155), an infinite set of nonseparable solutions of the two-dimensional Helmholtz equation (2.153) exists (Moseley 1965). Introducing the generative operator

$$\hat{L} = k_2 \frac{\partial}{\mathrm{d}k_1} - k_1 \frac{\partial}{\mathrm{d}k_2}, \tag{2.156}$$

one can find the nonseparable solution of $(n+1)$th order by means of the recursive relation

$$e_{n+1}(x, y) = \hat{L} e_n(x, y). \tag{2.157}$$

Substitution of the standard solution ($n = 0$), eqn (2.155), in eqn (2.157) yields simple examples of spatially nonseparable solutions of eqn (2.153):

$$e_1 = k_2 x \cdot \sin(k_1 x) \cdot \cos(k_2 y) - k_1 y \cdot \cos(k_1 x) \cdot \sin(k_2 y), \tag{2.158}$$

$$\begin{aligned} e_2 = {} & [(k_1 x)^2 + (k_2 y)^2] \cdot \cos(k_1 x) \cdot \cos(k_2 y) + k_2 y \cdot \cos(k_1 x) \cdot \sin(k_2 y) \\ & + k_1 x \cdot \sin(k_1 x) \cdot \cos(k_2 y) - 2 k_1 k_2 x y \cdot \sin(k_1 x) \cdot \sin(k_2 y). \end{aligned} \tag{2.159}$$

Zero-amplitude E_z, the customary condition for the boundary of a perfectly conducting waveguide, occurs at all points (x, y) that satisfy the two-dimensional transcendental equation

$$e_n = 0. \tag{2.160}$$

This condition is fulfilled for the familiar solution (2.155) ($n = 0$) on the rectangular contour (2.154); however, except for $n = 0$ none of the individual solutions e_n defines the close areas.

However, the solution e_0 perturbed by adding some proportion of the solution e_n produces a closed boundary if n is even. This boundary can be defined by an equation containing one constant C; thus for $n = 2$ we obtain

$$e_0 + C \cdot e_2 = 0. \tag{2.161}$$

Subject of C, various forms of waveguide crosssection can be found (Fig. 2.13). The trapped TM_{11} mode in the waveguide, shown in Fig. 2.13, is determined by the axial electric field distribution

$$E_z = E_0 \{e_0 + C \cdot e_2\} \exp[\mathrm{i}(\beta z - \omega t)]. \tag{2.162}$$

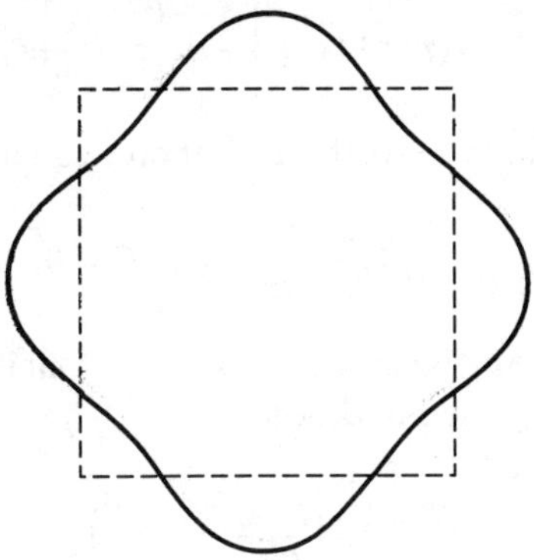

Fig. 2.13 Square-shape waveguide supporting the separable e_0 field of TM_{11} mode (broken line), and a waveguide with rounded corners and concave sides (solid line) supporting the combination of separable and non-separable TM_{11} modes (eqn 2.162), $C = 0.04$.

The other field components are derived from eqn (2.162) in the usual manner.

The analysis of boundary contour (2.161) shows that for $C > 0.01$ this combinatian results in a cross section that departs completely from the original. The maximum value of parameter C providing the formation of a closed boundary is $C = 0.0405$ (Moseley 1965). The perturbations of the cutoff frequencies of the waveguide $\Delta\omega_{cr}$ dictated by parameter C (eqn 2.161) depend on the ratio a/b (eqn 2.154). The maximum perturbation ($\Delta\omega_{cr}/\omega_{cr} \approx 0.12$) is achieved when $a = b$ (Schoonaert and Luypaert 1973). These nonseparable modes may be of interest for the design of waveguides with rounded corners and complex cross sections.

Unlike these stationary CW modes trapped in a waveguide, the nonstationary modulated weakly divergent wave beams in bulk plasma may be circumscribed by time-dependent nonseparable solutions of the KG equation. Let us start from the KG equation for such beams, using the cylindrical coordinate system (ρ, φ, z) with the z-axis along the direction of propagation:

$$\frac{\partial^2\Psi}{\partial\rho^2} + \frac{1}{\rho}\frac{\partial\Psi}{\partial\rho} + \frac{\partial^2\Psi}{\partial z^2} + \frac{1}{\rho^2}\frac{\partial^2\Psi}{\partial\varphi^2} - \frac{1}{c^2}\frac{\partial^2\Psi}{\partial t^2} = \frac{\Omega_e^2}{c^2}\Psi. \tag{2.163}$$

Here Ψ is one of the field components and Ω_e is the Langmuir plasma frequency. It is worth introducing a travelling coordinate

$$\tau = \frac{z - v_g t}{\sqrt{(1 - v_g^2 c^{-2}}}, \tag{2.164}$$

where v_g is the group velocity. The ansatz

$$\Psi = G(\rho, z) \cdot \exp\left[-\mathrm{i}\alpha\left(z - \frac{c^2}{v_{\mathrm{g}}}t\right)\right] \tag{2.165}$$

reduces eqn (2.163) directly to the Helmholtz equation

$$\nabla_{\perp} G + \frac{\partial^2 G}{\partial \tau^2} = k_{\perp}^2 G = 0. \tag{2.166}$$

Parameters α and $k_{\perp}$ may be interpreted as the longitudinal and transverse wavenumbers, obeying the condition

$$k_{\perp}^2 = \alpha^2\left(\frac{c^2}{v_{\mathrm{g}}^2} - 1\right) - \frac{\Omega_{\mathrm{e}}^2}{c^2}. \tag{2.167}$$

Introducing the new variable

$$\varepsilon = \sqrt{\rho^2 + \tau^2}, \tag{2.168}$$

one can write the exact localized solution of the Helmholtz equation (2.166) in the form (Ziolkowski *et al.* 1991)

$$G = \frac{I_{l+1/2}}{\sqrt{\varepsilon}}(\varepsilon) P_m^{\,l}(\tau/\varepsilon) \cdot \cos(m\phi), \quad l = 0, 1, 2, 3, \ldots \tag{2.169}$$

Here $I_{l+1/2}$ and P_l^m are Bessel function and Legendre associated function respectively. Substituting eqn (2.169) in eqn (2.165), one can see that as well as harmonic oscillations, the field Ψ at each point (ρ, φ, z) possesses some additional temporal variations described by the dependence of function G (eqn 2.169) on the coordinate τ; for example, the field Ψ on the beam axis is localized in the vicinity of the point $\tau = 0$.

Thus the numerous exact analytical solutions of Klein–Gordon and Helmholtz equations are broadened by different combinations of separable and nonseparable solutions. These fundamental equations are known to govern many basic wave phenomena in real media, optical and microwave systems, and cross-disciplinary physics. An example of the use of such generality for modelling nonstationary fields in a capillary cryoacoustical system will be given below in Section 3.6.

2.7 Conclusion: the genealogy of the Klein–Gordon equation

In discussing the Klein–Gordon (KG) equation, it should be remembered that eqn (2.12) became an effective instrument in classical mechanics about two centuries ago, long before the time when Klein and Gordon pioneered the use of a generalized version of this equation in quantum theory. Probably one of the first mentions of eqn (2.12) is connected with the theory of a string, which is considered to be one of the cornerstones of mathematical physics; a flexible string embedded in an elastic medium is described by the equation

$$\frac{\partial^2 F}{\partial z^2} - \frac{\rho_0}{T}\frac{\partial^2 F}{\partial t^2} = \frac{K_0}{T} F. \tag{2.170}$$

Here F is the displacement of the string in the z-direction, ρ_0 and T denote the linear density and tension and K_0 is a constant depending on the elastic properties of the string and its support.

Later, eqn (2.170) found a new lease of life in several fields of physics:

1. The development of quantum physics provided new interpretations of equations similar to (2.170). The important step was undertaken as a result of attempts to advance the relativistic generalization of the Schrödinger equation. Klein and Gordon suggested in the year 1927 the equation determining the wave function of a free relativistic charged particle with zero spin, and this equation has been named in their honour.

The KG equation is invariant under the Lorentz transformation of the independent variables (τ, η) to $(\tau', \eta',)$:

$$\begin{cases} \tau \to \tau' = \dfrac{\tau - v\eta}{\sqrt{(1-v^2)}}, \\ \eta \to \eta' = \dfrac{\eta - v\tau}{\sqrt{(1-v^2)}}, \end{cases} \tag{2.171}$$

where v is an arbitrary dimensionless velocity of propagation. Substitution of eqn (2.171) in eqn (2.12) leaves the form invariant, so the solutions of eqn (2.12) behave properly with regard to the space–time rotations of special relativity.

Using the quantum spatial (l_0) and temporal (t_0) scales for a particle with rest mass m,

$$l_0 = ct_0, \quad t_0 = \hbar(m_0 c^2)^{-1}, \tag{2.172}$$

and normalized variables $\eta = z \cdot l_0^{-1}$, $\tau = t \cdot t_0^{-1}$, one can reduce the equation discussed to the dimensionless form (2.12). Although such a spinless particle was unknown at that time, the Klein–Gordon equation was used later for the description of π-mesons (Schweber *et al.* 1955). As regards relativistic particles with spin, some echo of the KG equation may be discerned in the Dirac equation too. This equation, determining in particular one of the pair of spinor functions χ by means of operator ∇ and Pauli matrices $\hat{\sigma}$, may be written by means of scales (2.172) in the form, generalizing eqn (2.12),

$$(\hat{\sigma}\nabla)(\hat{\sigma}\nabla)\chi - \frac{\partial^2 \chi}{\partial \tau^2} = \chi. \tag{2.173}$$

Substitution of the plane wave $\chi = \exp[i\hbar(p\eta - \varepsilon\tau)]$ describing a free particle with energy ε and impulse p in eqn (2.173) yields the well-known relativistic result

$$p^2c^2 = \varepsilon^2 - (m_0c^2)^2. \tag{2.174}$$

This result shows a resemblance to the formulae (2.9) and (2.146) for the dispersion of EM fields in ionized gases and waveguides, derived from the Maxwell equations. These new fields of quantum and classical physics arising in the 1920s and 1930s opened up new horizons for the KG equation.

2. The development of electrical engineering and the design of transmission and communication lines drew attention to an equation similar to eqn (2.135) which was named the 'telegraph equation' by A. Poincaré as far back as 1897. This widely known equation, which can be reduced to the Klein–Gordon equation with 'imaginary mass', eqn (2.116), is considered so far as a fruitful model of numerous realistic electromagnetic systems (Wait 1985) and relevant wave systems in cross-disciplinary physics. Thus an unexpected application of this model to the random-walk problem in statistical physics, based on extension of the Fokker–Planck equation to phase space, leads to a telegraph equation in configuration space, determining the probability density ρ:

$$\frac{\partial^2 \rho}{\partial z^2} - \frac{1}{v^2}\frac{\partial^2 \rho}{\partial t^2} = \frac{1}{v^2 T}\frac{\partial \rho}{\partial t}. \tag{2.175}$$

Here v is the thermal velocity of the particles and T is a time constant, related to the viscosity of the medium (Davies 1954).

3. Progress in research on ultrafast power-dependent processes in microelectronics and optics is associated with the use of several types of nonlinear Klein–Gordon equations. Thus the distributed Josephson function is described by the so-called sine–Gordon (SG) equation (Scott 1969):

$$\frac{\partial^2 f}{\partial \eta^2} - \frac{\partial^2 f}{\partial \tau^2} = \sin f. \tag{2.176}$$

This equation is known to have both solitary and periodic exact solutions, expressed by the functions of the 'moving' coordinate $\xi = \eta - v\tau$, v being the dimensionless velocity of moving. The linear KG equation (2.12) relates to small values of the function $f(|f| \ll 1)$ in the SG equation (2.176). Taking into account both linear and cubic terms in the expansion of $\sin f$ in eqn (2.176) by the power series, one can obtain the nonlinear KG equation describing the stratified structures which correspond to the phase transitions in ferroelectric and ferromagnetic materials (Fei *et al.* 1992).

Novel applications of the KG and telegraph equations are associated with the exactly solvable model in the time-domain electrodynamics of ultrashort nonsinusoidal waveforms interacting with dispersive and conducting media (Chapter 3).

3

TIME-DOMAIN OPTICS OF TRANSIENTS IN DISPERSIVE AND ABSORBING MEDIA

3.1 Introduction: boundary conditions in the electrodynamics of nonsinusoidal fields

This chapter is devoted to an analysis of reflection, refraction and absorption phenomena in the processes of interaction of ultrashort arbitrarily shaped localized waveforms with dispersive and lossy media. Compared with traditionally considered solitary pulses containing many oscillations of an EM field, the waveforms discussed, containing only one or few field oscillations, are distinguished by the following features:

(1) the nonsinusoidal envelopes of the pulses;

(2) the considerable spectral bandwidth, comparable with the carrier frequency of the signal;

(3) propagation of transient signals without any carrier frequency.

These features hinder the use of the traditional Fourier method for analysis of nonstationary interactions of these waveforms with continuous media. In problems involving anharmonic excitation, a standard procedure to obtain the time-domain response is to calculate the frequency response and to use the numerical techniques of the inverse Fourier transform. However, these techniques have inherent difficulties with numerical instability, interpolation error, accumulation of errors with frequency marching, etc. (Sun and Rusch 1994). An additional source of difficulty is connected with the need for extensive computer memory to solve problems involving complex field envelopes. To avold these obstacles, an analytical approach to optics of transients is developed below without using Fourier 'ω–k' language. Being combined with the time-dependent presentations of EM fields in continuous media (Chapter 2), this approach forms an exactly solvable model in the time-domain optics of nonsinusoidal, transient, and single-cycle waveforms.

Considering the plane wave transients incident from free space on an interface of continuous medium, we should examine three models of such solitary waveforms:

1. To represent the smoothed waveform without a fixed edge, the Hermitian functions $\Phi_m(\xi)$ (Fig. 1.9) are of use. Since these functions are defined in the whole time range $(-\infty < t < \infty)$, one can express the incident waveform as a sum depending on the dimensionless travelling coordinate ξ:

$$\left.\begin{array}{l} E_{\text{in}}(\xi_-) \\ H_{\text{in}}(\xi_-) \end{array}\right\} = E_0 U(\xi_-), \quad \xi_\pm = \frac{t \pm z/c}{t_0}, \tag{3.1}$$

$$U(\xi_-) = \sum_{m=0}^{\infty} A_m \Phi_m(\xi_-). \tag{3.2}$$

Here t_0 is the time scale of an incident waveform, and the coefficients A_m are known to be determined by the orthogonality of the Hermitian functions Φ_m (eqn 1.32):

$$a_m = \int_{-\infty}^{\infty} U(\xi_-)\Phi_m(\xi_-)\,\mathrm{d}\xi. \tag{3.3}$$

The reflected waveform can be represented by means of the same functions Φ_m. Introducing the reflection coefficient R_m for the mth Hermitian harmonic Φ_m, we can write the components of the reflected field in the form

$$\left.\begin{array}{l} E_{\text{ref}}(\xi_+) \\ H_{\text{ref}}(\xi_+) \end{array}\right\} = E_0 \sum_{m=0}^{\infty} A_m R_m \Phi_m(\xi_+), \tag{3.4}$$

where ξ_+ is given by eqn (3.1).

2. To present a waveform with fixed zero-crossing, e.g. at the leading edge $(t = 0)$, defined at the half-limited time range $(t = 0)$, the Laguerre functions $L_m(\xi)$ prove to be suitable:

$$L_m(\xi) = \frac{e^{\xi/2}}{m!}\frac{\mathrm{d}^m}{\mathrm{d}\xi^m}(\mathrm{e}^{-\xi}\cdot\xi^m). \tag{3.5}$$

The travelling coordinate ξ is given by eqn (3.1). The functions L_m are orthonormal in the range $(0 \leqslant t < \infty)$:

$$\frac{1}{t_0}\int_0^{\infty} L_m\left(\frac{t}{t_0}\right)\cdot L_n\left(\frac{t}{t_0}\right)\mathrm{d}t = \delta_{mn}. \tag{3.6}$$

The envelopes of the waveforms, determined in the range $(0 \leqslant t < \infty)$, can be expanded using the property (3.6) to the series of Laguerre functions. The function L_0 is suitable for modelling asymmetric single-sided exponential pulses, excited for example photoconductive switches in coplanar transmission lines (Qian and Yamashita 1991) (Fig. 3.1). The linear combinations

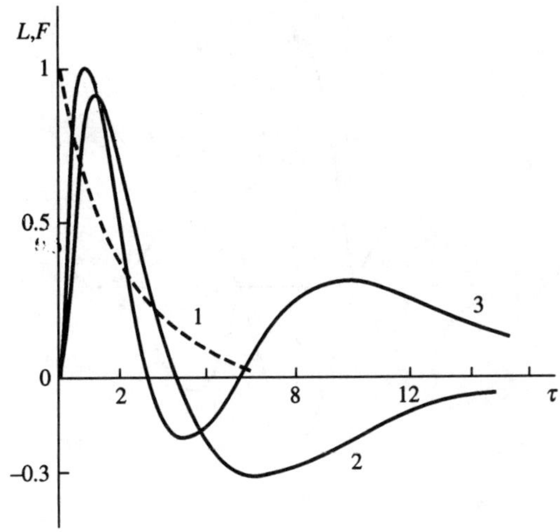

Fig. 3.1 Laguerre pulses vs. normalized time τ. Curves 1, 2, and 3 correspond to the functions L_0 (eqn 3.5), F_2 and F_3 (eqn 3.7) respectively.

$$F_m(\xi) = L_m(\xi) - L_{m+2}(\xi) \tag{3.7}$$

describe the envelopes with $m + 2$ zero-crossing points. One of these points is the pulse edge $\xi = 0$, the rate of growth of the envelope at this point is finite (Fig. 3.1). Single-cycle pulses similar to eqn (3.7) may be generated by the devices of terahertz optics (Fattinger and Grischkowsky 1989).

3. A significant group of solitary pulses is associated with wavelet signals. Being considered initially as a method of processing of information competing with Fourier analysis (Daubechies 1988), these signals are now of interest for the design of all-optical information circuits (Sheng *et al.* 1992) and remote sensing systems (Kaiser 1994). The wavelets are known to be generated by dilation and translation from a so-called mother wavelet $h(t)$:

$$h_{a,b}(t) = \frac{1}{\sqrt{a}} h\left(\frac{t-b}{a}\right) \tag{3.8}$$

with a dilation factor $a > 0$. The wavelet transform (WT) of the signal $S(t)$ is defined as the inner product in Hilbert space,

$$W_s(a, b) = \frac{1}{\sqrt{a}} \int_{-\infty}^{\infty} h^*\left(\frac{t-b}{a}\right) S(t)\, dt. \tag{3.9}$$

The $W_s(a, b)$ can be considered as functions of time shift b for each fixed

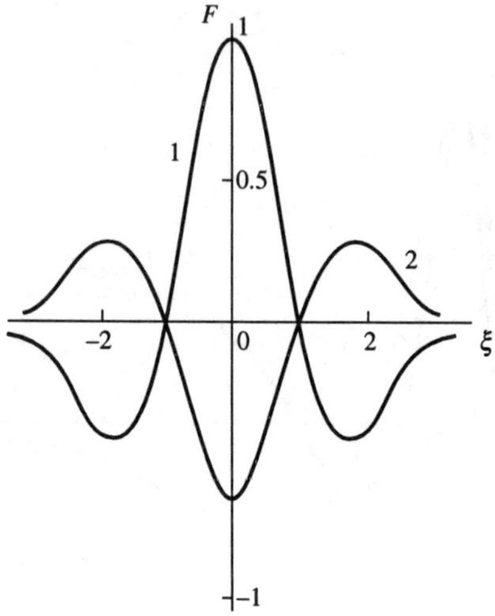

Fig. 3.2 'Mexican hat' wavelet envelopes vs. normalized time ξ. The normally incident and reflected waveforms (curves 1 and 2 respectively) illustrate the shape-preserving reflection of even Hermitian pulses obeying condition (3.10).

scale a, which display the information $S(t)$ at various levels of resolution[1]. A widely used example of pulses (3.8) is the 'Mexican hat' wavelet (1.37), shown in Fig. 3.2.

An essential property of wavelet real envelopes $h(t)$ is given by the formula (Szu *et al.*, 1992)

$$\int_{-\infty}^{\infty} h(t)\,\mathrm{d}t = 0. \tag{3.10}$$

This means that the wavelet must be oscillated to have a zero mean. The 'Mexican hat' obeys requirement (3.10) as well as the odd Hermitian functions $\Phi_{2m+1}(\xi)$ (eqn 1.31). Although each individual Laguerre function $L_m(\xi)$ does not satisfy to condition (3.10), this condition is fulfilled for the pairs of Laguerre functions $F_m(\xi)$, eqn (3.7). The time-domain solutions of

[1] It is interesting that certain classical formulae of optics, containing integral convolutions, may be interpreted by analogy with eqn (3.9). Thus the far-zone field in Fresnel diffraction (eqn 1.156) may be considered as the WT (3.9) of some signal $S = \Psi(x, y)$, characterizing the field distribution at the aperture plane $z = 0$, and the wavelet exponential function, containing the translations ξ and η and the dilation factor $a = \sqrt{zk^{-1}}$, coinciding with the Fresnel diffraction scale (Onural 1993).

the Klein–Gordon equation describing the non-separable fields *e* (eqn 2.36), also obey the wavelet condition (3.10).

All these types of waveform attract attention in the design of systems of signal processing, remote sensing and fractal communications. The development of transmitters of such nonsinusoidal signals is a current problem. The use of new wire sensors for measuring EM pulses in the subnanosecond region (Shen *et al.* 1990) and the detection of non-Fourier optical signals by photomultiplier tubes (Stoyanov and Dreischuh 1992) have demonstrated the advantages of nonstandard approaches to these problems. However, the fundamental redesign of traditional optical systems for operation with ultrashort nonsinusoidal transients has only recently begun.

To analyse the reflection–refraction processes for all the above-mentioned models of transient waveforms (1)–(3) at the interfaces of continuous media, the time-domain boundary conditions must be used. In the simple case of normal incidence of an EM field these conditions are known to amount to the standard requirement of continuity of the electric and magnetic components at the plane boundary interface, $z = 0$. In the case of oblique incidence these conditions are different for the S- and P-polarizations of the incident transients. These continuity conditions are sufficient for the determination of time-dependent reflection coefficients for solitary waveforms, the polarization and angle of incidence being arbitrary.

The decisive influence of the shapes and time scales of anharmonic single-cycle waveforms on their interactions with continuous media is visualized here through the time-domain description of both incident pulses and nonstationary EM fields excited in these media. Some important remarks should be stressed in regard to such interactions:

1. Both pulses and continuous media are characterized by definite time scales; the reflectivity of the interface proves to depend on the ratio of these scales, other parameters being invariant. This effect may be considered as a generalization of the frequency dispersion of the reflection coefficient in Fourier optics.

2. The reflection of Hermitian waveforms from a given interface may be discerned as appreciably subject to the shapes of these waveforms, their time scales being equal. Such phenomena were neglected in the traditional electrodynamics of modulated harmonic pulses limited by the fixed sinusoidal profiles of field oscillations. This peculiar dispersion of the reflection of short Hermitian waveforms, depending on their shapes, promises rapid controlled reshaping of waveforms in real time.

3. Unlike the Hermitian pulses, the reflection of Laguerre signals from both dielectric and conducting media does not possess any shape-dependent dispersion. This shape-preserving reflection may be of interest for many ultrafast all-optical information systems.

4. Unlike the traditional representatlons of narrow pulses with steepened

fronts by means of δ-functions or unit step Heaviside functions, we consider analytically the waveforms with finite time scales and smooth envelopes.

5. Instead of complex integration in the frequency domain, the reflected pulses and refracted nonstationary fields are found in time-domain electrodynamics by solution of a linear system of algebraic equations. This approach results in a drastic simplification of the analysis of nonstationary interaction of EM pulses with continuous media.

The nonstationary refracted fields excited in reflecting media by ultrashort transients are considered here by means of the Debye model of dielectrics with free carriers (plasma-like media). Some properties of such fields were examined above in Chapter 2. This approach is developed below for the time-domain description of refracted fields in the materials widely used in integrated optics and microelectronics (semiconductors, ionic solids, transmission lines and metal films). Nonstationary EM fields may be generated by shock excitation of plasma-filled microwave sources, backward-wave oscillators, and submillimetre and microwave lasers. The closeness of the characteristic time scales of picosecond and femtosecond transients to the periods of the eigenfrequencies of such systems is shown to provide sensitivity of reflection–refraction processes to the shapes of nonsinusoidal and single-cycle transients.

Such phenomena are specified below for different envelopes of single-cycle pulses, their polarization and geometry of incidence, and dielectric and conductive reflecting materials. Controlled shaping and polarization of transients may be of interest for optimization of systems of ultrahigh-speed fractal communications. Sorting of arbitrary nonsinusoidal waveforms in real time, based on difference in their reflection coefficients subject to time scales, shapes and polarization, opens up new possibilities for all-optical processing of light signals. Complementing the technique of sub-femtosecond waveform measurements based on frequency-resolved optical gating, this approach may be of interest for extension of holographic processing into the femtosecond time domain (Weiner *et al.* 1992).

The physical fundamentals of reflection, refraction, and absorption phenomena in the time-domain optics of transients are developed in Section 3.2. The procedure for direct analytical determination of reflection and transmission coefficients for arbitrary waveforms beyond the scope of Fourier expansions is illustrated by means of Hermitian pulses interacting with dispersive and lossy media. The dependence of these interactions upon the pulse shapes is emphasized.

The special properties of Laguerre transients, possessing shape-independent reflection in both dielectric and conducting media, are demonstrated in Section 3.3. Using this feature, an exactly solvable model for these processes is advanced. The reflection coefficients for waveforms containing arbitrary combinations of Laguerre transients are represented by explicit algebraic formulae.

Polarization phenomena in the optics of transients are illustrated in Section 3.4 by means of the difference in reflection coefficients of S- and P-polarized waveforms incident obliquely on a dispersive interface. The shape-dependent Brewster effect for P-polarized transients is examined. This approach visualizes the effect of shape-dependent selection of waveforms reflected from a dispersive plane interface.

Nonstationary refracted fields excited by ultrashort transients in dispersive media show the influence of the shapes of incident Hermitian waveforms on the angles of refraction (Section 3.5). Rapid spreading of transmitted pulses propagating in such media is examined by means of time-domain solutions; however, the possibility of preservation of some parameters of these fields is illustrated.

The sensitivity of the dynamics of ultrashort solitary pulses interacting with transparent and conducting media to their envelopes stimulates the development of devices for flexible modelling of these phenomena. The possibilities of using cryoacoustical capillary waveguides and transmission lines for modelling of the relevant pulse effects are considered in Section 3.6.

The physical phenomena and computational advantages revealed by exactly solvable models in time-domain optics are considered in the final Section 3.7.

3.2 Hermitian pulse optics

This section is intended to illustrate the shape-dependent phenomena in the reflection of Hermitian pulses from the interface of lossless dielectric media as well as in absorption in lossy media. The procedure for analysis of these phenomena may be considered as a generalization of the approach well known in frequency-domain electrodynamics. Defining the reflection coefficient R as the ratio of the complex amplitudes of the reflected E_{ref} and incident E_0 electric components of a monochromatic EM field, $R = E_{\text{ref}}E_0^{-1}$ and restricting ourselves first to the simple case of normal incidence of a linearly polarized field, we can write one boundary condition at the interface, $z = 0$:

$$E_0(1 + R) = E_{\text{t}}. \tag{3.11}$$

Determining the field's magnetic components H from Maxwell equation (1.44) and imposing the continuity of components H, we obtain another boundary condition at the plane $z = 0$:

$$E_0(1 - R) = H_{\text{t}}. \tag{3.12}$$

Here E_{t} and H_{t} are the electric and magnetic components of the transmitted EM field. Introducing the transmission coefficients T_1 and T_2 for the monochromatic wave,

$$E_t = T_1 E_0, \quad H_t = T_2 E_0, \tag{3.13}$$

one can find from eqns (3.11)–(3.13) that

$$\frac{1+R}{1-R} = \frac{T_1}{T_2}. \tag{3.14}$$

Equation (3.14) is the corollary of the boundary conditions. Use of the refractive index n leads to the following correlation between the transmission coefficients T_2 and T_1 (eqn 3.13): $T_2 = nT_1$. Substitution of this correlation in eqn (3.14) yields the Fresnel formula for the reflection coefficient of a monochromatic wave:

$$R = \frac{1-n}{1+n}. \tag{3.15}$$

Since the refractive index n is known to depend on the frequency of the wave, the reflection coefficient R and transmission coefficients T_1, T_2 (eqn 3.13) are also frequency-dependent.

Proceeding in a similar fashion, we examine in Subsection (3.2.1) the reflection of nonsinusoidal Hermitian pulses from a dielectric interface.

As regards shape-dependent attenuation of these pulses in lossy dielectrics, one can use the model of a Debye medium. Thus, considering a dielectric with free carriers and calculating the Joule losses associated with collisions of these carriers, we can begin with a general formula, well known for example in plasma electrodynamics:

$$G = \int_{-\infty}^{\infty} \boldsymbol{J}_\omega \boldsymbol{E}_\omega^* \, d\omega. \tag{3.16}$$

Here $\boldsymbol{J}_\omega$ is the Fourier transform of the current, induced by an electric field $\boldsymbol{E}_\omega$, and G is the energy absorbed in unit volume. This formula is used in Subsection 3.2.2. for the analysis of shape-dependent absorption of Hermitian pulses in a Debye medium.

3.2.1 Reflection of Hermitian waveforms from transparent dielectrics

The continuity of the components of an EM field at a dielectric interface is a cornerstone for both frequency-domain and time-domain analysis of reflection–refraction phenomena. To find the coefficients of reflection R and transmission T_1, T_2 for Hermitian pulses, one can expand these nonstationary incident and reflected fields at the interface $z = 0$ to the series of orthonormal Hermitian functions Φ_m, eqn (3.4). As regard the nonstationary transmitted fields, their electric E_t and magnetic H_t components may be represented by means of the nonseparable fields analysed in

Chapter 2. Making use of expressions for these fields at the plane $z=0$ (eqn 2.38), we can write the components E_t and H_t in the form

$$\left.\begin{matrix} E_t(\tau) \\ H_t(\tau) \end{matrix}\right|_{z=0} = -\frac{E_0}{2}\sum_{q=1}^{\infty} a_q[J_{q-1}(\tau) \mp J_{q+1}(\tau)]. \tag{3.17}$$

Expansion of transmitted fields (3.17) in terms of Hermitian functions ϕ_m results in the equations (Shvartsburg and Stenflo 1994)

$$\left.\begin{aligned} E_t(\tau)\,|_{z=0} &= -E_0 \sum_{m=0}^{\infty} T_{1m}\Phi_m\left(\frac{t}{t_0}\right), \\ H_t(\tau)\,|_{z=0} &= -E_0 \sum_{m=0}^{\infty} T_{2m}\Phi_m\left(\frac{t}{t_0}\right). \end{aligned}\right\} \tag{3.18}$$

The coefficients $T_{1,2m}$ in eqns (3.18), generalizing the steady-state transmission coefficients $T_{1,2}$ introduced above (eqn 3.12) for the description of monochromatic wave trains, are related to the coefficients a_q by means of matrices P_{mq} and Q_{mq}. These matrices describe the excitation of the qth harmonics of the electric and magnetic components of the nonseparable EM field (eqn 3.17) by the mth incident Hermitian pulse:

$$T_{1m} = \sum_{q=1}^{\infty} a_q P_{mq}(\Omega_e T_0), \tag{3.19}$$

$$T_{2m} = \sum_{q=1}^{\infty} a_q Q_{mq}(\Omega_e T_0), \tag{3.20}$$

where

$$\left.\begin{matrix} P_{mq}(\Omega_e t_0) \\ Q_{mq}(\Omega_e t_0) \end{matrix}\right\} = \frac{1}{2t_0}\int_{-\infty}^{\infty} \Phi_m\left(\frac{t}{t_0}\right)[J_{q-1}(\Omega_e t) \mp J_{q+1}(\Omega_e t)]\,dt. \tag{3.21}$$

Thus each of the matrix elements P_{mq} and Q_{mq} (3.21) depends only on the ratio of the time scales of the pulse t_0 and the dielectric medium Ω_e^{-1}, the numbers of the exciting pulse m and excited harmonic q being fixed.

For simplicity, the high-frequency dielectric permittivity ε_∞ is assumed to be equal to unity here.

The subsequent analysis of the reflection–refraction problem contains three steps:

1. The coupled determination of the reflection coefficients R_m and the structure of the transmitted field (the coefficients a_q in the field representation (2.36)).

Using the expansion of the incident (3.2), reflected (3.4) and transmitted (3.18) fields in terms of orthonormal functions Φ_m, one can introduce the reflection R_m and transmission $T_{1,2m}$ coefficients for mth Hermitian pulse. The standard boundary conditions yield the system of algebraic equations

$$1 + R_m = T_{1m}, \quad 1 - R_m = T_{2m}. \tag{3.22}$$

To determine R_m we take into account that:

(a) each value of the positive number m corresponds to a pair of equations (3.22);

(b) the nonzero elements of the matrices P_{mq} and Q_{mq} have indices of different parity (thus $P_{11} = 0$ but $P_{12} \neq 0$, etc.);

(c) if the incident waveform at $z = 0$ is described by an odd function, i.e. if

$$U\left(\frac{t}{t_0}\right)\bigg|_{z=0} = -U\left(-\frac{t}{t_0}\right)\bigg|_{z=0}, \tag{3.23}$$

then the even Hermitian functions do not appear in the representation of the incident envelope; in this case the pair of equations that corresponds to even indices m is reduced to

$$T_{1m} = T_{2m} = 0. \tag{3.24}$$

As an example, we now focus attention on the reflection of a normally incident pulse that is described only by the Hermitian function Φ_1. Making use of eqns (3.22), we can write

$$\left.\begin{aligned} 1 + R_1 &= a_2P_{12} + a_4P_{14} + a_6P_{16} + \ldots, \\ 1 - R_1 &= a_2Q_{12} + a_4Q_{14} + a_6Q_{16} + \ldots \end{aligned}\right\} \tag{3.25}$$

since Φ_3, Φ_5 and other Φ_{2n-1} pulses are absent in the incident waveform, we use condition (3.24), i.e.

$$\left.\begin{aligned} 0 &= a_2P_{32} + a_4P_{34} + a_6P_{36} + \ldots, \\ 0 &= a_2Q_{32} + a_4Q_{34} + a_6Q_{36} + \ldots \end{aligned}\right\} \tag{3.26}$$

Truncating the system of equations (3.25)–(3.26), i.e. considering l pairs of such equations, one can calculate the reflection coefficient R_1 and $2l - 1$ coefficients a_q that determine the fields excited inside the medium.

The time-domain reflection coefficient R_m can be represented owing to eqn (3.22) in a form similar to eqn (3.15):

$$R_m(\alpha) = \frac{1 - T_{2m}(\alpha) \cdot T_{1m}^{-1}(\alpha)}{1 + T_{2m}(\alpha) \cdot T_{1m}^{-1}(\alpha)}, \quad \alpha = \Omega_e t_0 \tag{3.27}$$

Thus the ratio $T_{2m}T_{1m}^{-1}$ in the time-domain approach is reminiscent of the refractive index $n(\omega)$ in the frequency-domain formula (3.15):

$$\frac{T_{2m}(\alpha)}{T_{1m}(\alpha)} \to n(\omega). \tag{3.28}$$

However, besides the dependence of R_m on the pulse time scale t_0, this reflection coefficient depends on the shape of the waveform, determined by number m, the time scale t_0 being given.

To calculate the coefficients $R_{m.}$, T_{1m}, T_{2m} we have to find the matrix elements P_{mq} and Q_{mq} (3.21).

2. The analytical calculation of matrix elements $P_{mq}(\alpha)$ and $Q_{mq}(\alpha)$. It is significant that the matrix elements (3.21) can be found in a general form analytically by means of an integral (Ryzhik and Gradstein 1980):

$$\int_0^\infty J_{2\mathrm{p}}(\alpha x)\exp\left(-\frac{x^2}{2}\right)\mathrm{d}x = \sqrt{\pi/2}\,\exp\left(-\frac{\alpha^2}{4}\right) I_\mathrm{p}\left(\frac{\alpha^2}{4}\right), \qquad (3.29)$$

where I_p is the modified Bessel function.

The simplest matrix elements $P_{mq}(\alpha)$ and $Q_{mq}(\alpha)$ corresponding to the zeroth-order Hermitian envelope $\Phi_0 = C_0\exp(-t^2/(2t_0^2))$ can be obtained readily from expression (3.28):

$$\left.\begin{aligned} &\left.\begin{matrix} P_{0q}(\alpha) \\ Q_{0q}(\alpha) \end{matrix}\right\} = C_0 K\left[I_{\frac{q-1}{2}}(\alpha^2/4) \mp I_{\frac{q-1}{2}}(\alpha^2/4)\right] \\ &K = \sqrt{\pi/2}\,\exp(-\alpha^2/4), \quad C_0 = \pi^{-1/4},\ \alpha = \Omega_\mathrm{e} t_0. \end{aligned}\right\} \qquad (3.30)$$

Manipulations with the convolutions (3.21) result in more complicated formulae for Hermitian pulses of higher order $(m > 0)$. i.e.

$$\left.\begin{aligned} P_{1q}(\alpha) &= \frac{C_0 K\alpha}{\sqrt{2}}\left[I_{\frac{q-2}{2}}\left(\frac{\alpha^2}{4}\right) - 2\cdot I_{\frac{q}{2}}\left(\frac{\alpha^2}{4}\right) + I_{\frac{q+2}{2}}\left(\frac{\alpha^2}{4}\right)\right], \\ Q_{1q}(\alpha) &= \frac{C_0 K\alpha}{\sqrt{2}}\left[I_{\frac{q-2}{2}}\left(\frac{\alpha^2}{4}\right) - I_{\frac{q+2}{2}}\left(\frac{\alpha^2}{4}\right)\right], \end{aligned}\right\} \qquad (3.31)$$

$$P_{21} = \sqrt{2}\,C_0 K\left[(1 - 2\alpha^2)I_0\left(\frac{\alpha^2}{4}\right) + (3 + 2\alpha^2)I_1\left(\frac{\alpha^2}{4}\right)\right], \qquad (3.32)$$

$$Q_{21} = \sqrt{2}\,C_0 K\left[I_0\left(\frac{\alpha^2}{4}\right) - 3I_1\left(\frac{\alpha^2}{4}\right)\right]. \qquad (3.33)$$

The factor K is defined in eqn (3.30). Proceeding in a similar fashion, one can find the matrix elements (3.21) with arbitrary indices m and q. To simplify the calculations, some values of the functions $P_{mq}(\alpha)$ and $Q_{mq}(\alpha)$ are tabulated in Appendix A.3.

Shape-dependent reflection of Hermitian pulses Φ_1 and Φ_3 is illustrated in Figs 3.3 and 3.4. The reflection coefficients differ for these incident pulses even if they are characterized by the same dimensionless duration $\alpha = \Omega_\mathrm{e} t_0$. As an example, when $\alpha = 5$, $R_1(5) = -0.72$ and $R_3(5) = -0.23$; this means that the amounts of energy projected to the medium by pulses Φ_3 and Φ_1, $W_m = 1 - R_m^2$, in this case are in the ratio 2:1.

3. Time-domain determination of transmitted fields.

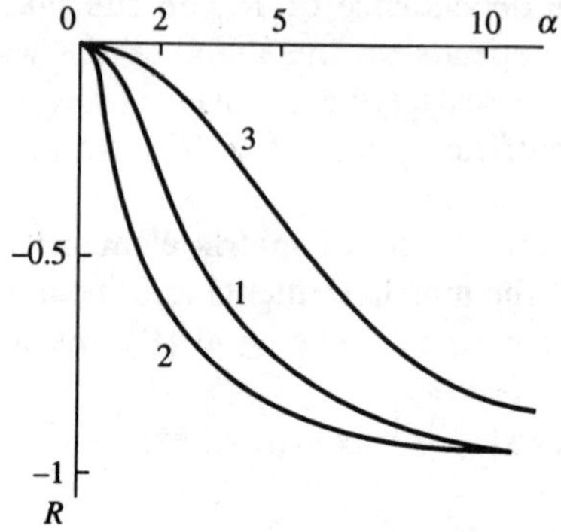

Fig. 3.3 Shape-dependent coefficients of reflection R of normally incident Hermitian waveforms Φ_1 and Φ_3 (curves 1 and 3 respectively) and Laguerre waveforms (curve 2) from a dielectric with free carriers, vs. the normalized time scale $\alpha = \Omega_e t_0$.

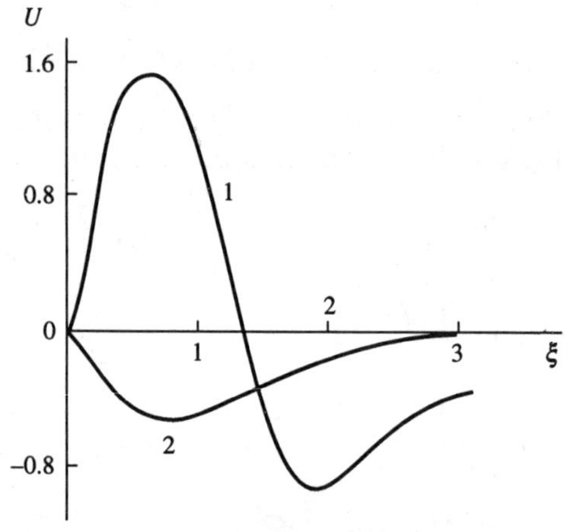

Fig. 3.4 Shape-dependent reflection of a normally incident dimensionless waveform U containing the Hermitian pulses Φ_1 and Φ_3: $U = \Phi_1 - 2\Phi_3$. Curves 1 and 2 show the incident and reflected envelopes respectively, $\alpha = 2$. The difference in the reflection coefficients R_1 and R_3 (Fig. 3.3) results in strong distortion of the reflected envelope.

The coefficients a_q determining the excitation of the qth harmonics e_q and h_q (eqn 2.36) in a dispersive medium can be found from the system of linear algebraic equations (3.22). Attention may be drawn to some similarity between this procedure and the calculation of the excitation of waveguide modes by a pulse incident on the waveguide's input. However, unlike the waveguide modes, the functions $\Psi_q(\tau, \eta)$ are not orthogonal;

therefore an expansion of these functions at the interface $z=0$ in terms of orthonormal Hermitian pulses is required for the problem under discussion. It is useful to point out that even (odd) Hermitian pulses excite odd (even) nonseparable harmonics.

The nonstationary transmitted field at an arbitrary point within the refracting medium can be found by substitution of the coefficients a_q obtained in the representations of field components via the functions Ψ_q, eqns (2.36)–(2.37). Using the asymptotes of the functions Ψ_q (eqn 2.39), we obtain the simple expressions for the field at the wings of the transmitted pulses. Thus for example the dimensionless electric component (eqn 2.33) at the wings of the field excited by an odd antisymmetric Hermitian waveform may be written as a damping harmonic wave,

$$e|_{\tau^2 \gg \eta^2} = \sqrt{\frac{2}{\pi\tau}} \cos(\tau + \pi/4) \sum_{n=0}^{\infty} (-1)^n a_{2n+1}. \tag{3.34}$$

It is expedient to point out some properties of the reflection coefficients R_m corresponding to different Hermitian waveforms Φ_m:

(a) Unlike the reflection coefficient in Fourier optics, the time-domain coefficients R_m are always real (the phase of R_m is not defined).

(b) A negative reflection coefficient indicates that the polarization direction of the reflected pulse has changed to the opposite one;

(c) The matrices $P_{mq}(\alpha)$ and $Q_{mq}(\alpha)$ (eqn 3.21) depend on the product $\alpha = \Omega_e t_0$, which characterizes the time scale of the incident pulse with respect to the cutoff period of the dispersive medium. The matrix elements are continuous functions for all values of the variable $\alpha = \Omega_e t_0$, i.e. also at the value $\Omega_e t_0 = 2\pi$, which corresponds to total reflection of harmonic radiation with the period t_0. However, the reflection coefficient of Hermitian pulse Φ_1 is $R_1(2\pi) = -0.8$ (Fig. 3.3); this means that ~36% of the energy of this pulse $(1 - R_1^2 = 0.36)$ penetrates into the reflecting dielectric.

(d) The asymptotes of matrix elements $P_{mq}(\alpha)$ and $Q_{mq}(\alpha)$, related to long incident pulses or dense media ($\Omega_e t_0 \gg 1$), may be analysed by means of asymptotes of modified Bessel functions (A.2.4). Using this formula, we find from eqns (3.19)–(3.20) that

$$\lim \left. \frac{T_{1m}}{T_{2m}} \right|_{\alpha\to\infty} = \frac{\text{const.}}{\alpha^2}. \tag{3.35}$$

Combining eqn (3.35) with eqn (3.22), we can see that the values of the reflection coefficients $R_m(\alpha)$ in the range $\alpha \gg 1$ tend to the limit (Fig. 3.3)

$$\lim R_m(\alpha)|_{\alpha\to\infty} = -1. \tag{3.36}$$

(e) The reflection of Hermitian pulses in the case $\Omega_e t_0 \ll 1$ (short pulses or rarefied reflecting media) may be analysed by other asymptotes of modified Bessel functions (A.2.5). Using this formula, we obtain from eqns (3.19)–(3.20)

$$\lim \frac{T_{1m}}{T_{2m}}\bigg|_{\alpha\to 0} = 1 - C_{\mathrm{m}}\cdot\alpha^2. \tag{3.37}$$

Here C_m is some positive constant depending on the number of Hermitian pulse Φ_m. The combination of eqn (3.37) with eqn (3.22) yields another asymptotic property for reflection coefficients $R_{\mathrm{m}}(\alpha)$ (Fig. 3.3):

$$\lim R_m(\alpha)\,|_{\alpha\to 0} = -\tfrac{1}{2}\cdot C_m\cdot\alpha^2. \tag{3.38}$$

Since the model of a reflecting transparent dielectric with a density of free carriers N_{e} is discussed here, and $\alpha^2 = \Omega_{\mathrm{e}}^2\cdot t_0^2 \propto N_{\mathrm{e}}$ (eqn 1.200), one can see from eqn (3.38) that the reflectivity of this medium is proportional to the density of individual scatterers: $R_m \propto N_{\mathrm{e}}$. This reflectivity being weak ($|R_m(\alpha)| \ll 1$), we can examine below the Joule losses of Hermitian pulses in this medium, supposing the pulse shapes to be given.

3.2.2 Shape-dependent absorption of Hermitian waveforms in a Debye dielectric

In contrast to Section 1.7, which considered the absorption of Hermitian pulses by quantized systems of molecular terms, the discussion here is of Joule losses of such pulses propagating through a Debye medium. This purely classical problem may be solved by means of formula (3.16), Keeping in mind for example the Joule losses of the free-carrier current $\boldsymbol{J}$ in the dielectric, characterized by the frequency of collisions ν of the carriers, let us write the FT of this current:

$$\boldsymbol{J}_\omega = \frac{\Omega_{\mathrm{e}}^2}{4\pi}\cdot\frac{\nu\cdot\boldsymbol{E}_\omega}{\nu^2+\omega^2}. \tag{3.39}$$

Substitution of eqn (3.39) in eqn (3.16) determines the density of energy losses G:

$$G = \frac{\Omega_{\mathrm{e}}^2\cdot\nu}{4\pi}\int_{-\infty}^{\infty}\frac{|\boldsymbol{E}_\omega|^2\,\mathrm{d}\omega}{\nu^2+\omega^2}. \tag{3.40}$$

The shape of the absorption line in this medium, characterized by the factor $\nu(\omega^2+\nu^2)^{-1}$, has no resonant frequencies.

To examine the absorption of e.g. Hermitian waveform Φ_1 by means of eqn (3.40), one can use the FT of an envelope Φ_1, which was found above (eqn 1.35). Substituting this FT in eqn (3.40) and introducing the dimensionless variable $x = \omega\cdot t_0$, we obtain

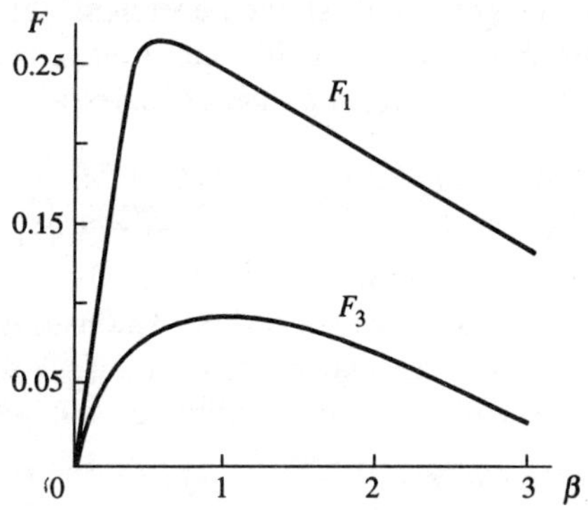

Fig. 3.5 Shape-dependent absorption of Hermitian waveforms in a Debye medium; the normalized losses per unit volume F (eqn 3.42) are plotted vs. the dimensionless collisional parameter $\beta = \nu T_0$. The curves F_1 and F_3 relate to absorption of Hermitian waveforms Φ_1 and Φ_3 (eqn 1.31) respectively.

$$G_1 = E_0^2(\Omega_e t_0)^2 F_1(\beta), \quad \beta = \nu t_0, \tag{3.41}$$

$$F_1(\beta) = \beta \int^{\infty} \frac{x^2 \exp(-x^2)\,\mathrm{d}x}{x^2 + \beta^2}. \tag{3.42}$$

Here β is the normalized breadth of the absorption line. Integral $F_1(\beta)$ is expressed in terms of the error function erf(β):

$$F_1(\beta) = \beta\{1 - \sqrt{\pi}\,\beta \exp(\beta^2)\,[1 - \mathrm{erf}(\beta)]\}. \tag{3.43}$$

Repeating the same procedure for the Hermitian waveform Φ_3 (eqn 1.31), we can write

$$E_\omega = -8E_0 t_0 C_3 \sqrt{\pi/2}\,\omega t_0 \exp\left[-\frac{(\omega t_0)^2}{2}\right][2(\omega t_0)^2 - 3]. \tag{3.44}$$

Substitution of eqn (3.44) in eqn (3.40) yields the result of integration in a form similar to eqn (3.41):

$$\left.\begin{aligned} G_3 &= E_0^2(\Omega_e t_0)^2 F_3(\beta), \\ F_3(\beta) &= \frac{\beta}{12}(3 + 2\beta^2)\{2(1 + \beta^2) - \sqrt{\pi}\,\beta(3 + 2\beta^2)\exp(\beta^2)\,[1 - \Phi(\beta)]\}. \end{aligned}\right\} \tag{3.45}$$

The graphs $F_1(\beta)$ and $F_3(\beta)$, shown in Fig. 3.5, illustrate the dependence of absorption of Hermitian pulses upon their normalized time scales β and shapes. The energy losses of these pulses prove to be different even when their time scales are equal (shape-dependent absorption). When the time scale t_0 or the frequency of collisions ν decreases so that the parameter β becomes small ($\beta \ll 1$), the energy losses also decrease:

$$\lim F_1|_{\beta\to 0} = \beta, \quad \lim F_3|_{\beta\to 0} = \beta/2. \tag{3.46}$$

Although the normalized energy losses, eqns (3.42) and (3.45), are calculated in the frequency domain, these exact analytical results do not contain any frequencies. Hence the absorption of nonsinusoidal waveforms in a classical Debye medium as well as in a quantum system (Section 1.7) possesses a strong dependence on the shape of the waveform.

3.3 Laguerre pulse optics

Laguerre waveforms $L_m(\xi)$ (Fig. 3.1) are finding steadily increasing applications for communication purposes (Papazoglou 1975), shaping of signals in terahertz optics (Frankel *et al.* 1989), and sounding of solitary targets (Ma and Ciric 1992). The' unique features of these waveforms, determining their important role in time-domain pulse optics, are:

(1) the fast rising edge, followed by a long exponentially decaying tail;

(2) the flexible rate of increase in the rising edge of the waveforms $F_m(\xi)$ (eqn 3.7), which may be shaped as ξ, ξ^2, ξ^3, and so on;

(3) the peculiar integral property of the Laguerre waveform,

$$\frac{1}{2}\int_0^\infty L_m(\xi)\,\mathrm{d}\xi = (-1)^m, \tag{3.47}$$

simplifying the selection of linear combinations of these waveforms, obeying condition (3.10).

Alongside these properties, the interest in Laguerre pulses is based on the explicit algebraic expressions for the reflection coefficients of these pulses in the time domain. These formulae reveal the monotonic dependence of the reflectivity of dielectric and conducting interfaces on the parameters $\alpha = \Omega_e t_0$ (eqn 3.27) and $\beta = t_0 T_\sigma^{-1}$ (eqn 2.111) respectively. The reflection coefficients prove to be equal for the arbitrary Laguerre waveforms characterized by fixed values of parameter α or β (Shyartsburg 1994). These results, forming an exactly solvable model of reflection of Laguerre pulses, are examined here for nonconducting (Subsection 3.3.1) and conducting (Subsection 3.3.2) materials.

3.3.1 Distortionless reflection of Laguerre waveforms from dielectrics with free carriers

The reflection of normally incident Laguerre waveforms $L_m(\xi)$, eqn (3.5) may be analysed in the time domain by analogy with Hermitian pulses. Using the continuity of the fields at the boundary interface, one can obtain the system of equations (3.22) with the transmission coefficients T_{1m} (eqn 3.19) and T_{2m} (eqn 3.20); however, the matrix elements $P_{mq}(\alpha)$ and $Q_{mq}(\alpha)$ (eqn 3.21) must now be replaced by new expressions $M_{mq}(\alpha)$ and $N_{mq}(\alpha)$:

$$\left.\begin{array}{l} P_{mq}(\alpha) \rightarrow M_{mq}(\alpha) \\ Q_{mq}(\alpha) \rightarrow N_{mq}(\alpha) \end{array}\right\} = \frac{1}{t_0}\int_0^\infty L_m\left(\frac{t}{t_0}\right)[J_{q-1}(\Omega_e t) \mp J_{q+1}(\Omega_e t)]\,\mathrm{d}t. \tag{3.48}$$

The combinations of Bessel functions in the convolution (3.48) associated with the boundary representations of nonstationary fields (eqn 3.17) indicate that the discussion is of the reflectivity of a plasma-like medium with free carriers.

It is worth introducing the new matrices $X_{mq}(\alpha)$ and $Y_{mq}(\alpha)$:

$$\left.\begin{array}{l} X_{mq}(\alpha) \\ Y_{mq}(\alpha) \end{array}\right\} = \frac{1}{2}[N_{mq}(\alpha) \mp M_{mq}(\alpha)]. \tag{3.49}$$

It follows from eqns (3.48)–(3.49) that

$$\left.\begin{array}{l} X_{mq}(\alpha) \\ Y_{mq}(\alpha) \end{array}\right\} = \int_0^\infty \mathrm{d}x\, L_m(x)\left\{\begin{array}{l} J_{q-1}(\alpha X) \\ J_{q+1}(\alpha X) \end{array}\right\}. \tag{3.50}$$

Making use of matrices $X_{mq(\alpha)}$ and $Y_{mq}(\alpha)$, one can rewrite eqns (3.22) in the form

$$\sum_{q=1}^{\infty} a_q Y_{mq}(\alpha) = -2, \quad \sum_{q=1}^{\infty} a_q X_{mq}(\alpha) = 2\cdot R_m. \tag{3.51}$$

To find the correlation between the matrix elements $X_{0q}(\alpha)$ and $Y_{0q}(\alpha)$, let us take advantage of the integral (Ryzhik and Gradstein 1980)

$$\int_0^\infty \mathrm{e}^{-px} J_{q+2}(\alpha x)\,\mathrm{d}x = \alpha^2 A^2 \int_0^\infty \mathrm{e}^{-px} J_q(\alpha x)\,\mathrm{d}x, \tag{3.52}$$

$$A = [p + \sqrt{(p^2+\alpha^2)}]^{-1}. \tag{3.53}$$

Calculating the integrals (3.50) by means of formulae (3.52)–(3.53), we obtain the simple correlation

$$X_{0q}(\alpha) = \alpha^2 A^2 Y_{0q}(a). \tag{3.54}$$

Substitution of eqn (3.54) in the system (3.51) yields an explicit expression for the reflection coefficient $R_m(\alpha)$. Using the function A (eqn 3.53) with the value of parameter $p = 0.5$, related to the convolution (3.50) containing the Laguerre function, we obtain finally an algebraic formula for the coefficient R in the time domain:

$$R(\alpha) = -\frac{4(\Omega_e t_0)^2}{\{1 + \sqrt{[1 + 4(\Omega_e t_0)^2]}\}^2}. \tag{3.55}$$

The monotonic dependence of reflection coefficient R on the parameter α is shown in Fig. 3.3, $\lim R|_{\alpha\to\infty} = -1$. As expected, long incident pulses and dense reflecting media ($\alpha \gg 1$) correspond to the case $\lim R|_{\alpha\to\infty} = -1$.

It is worth while to discuss this result together with a similar one, relating to the reflection of the same waveforms from the plane boundary of a conducting material.

3.3.2 Shape preservation of Laguerre waveforms reflected from a conducting interface

To consider the time-domain reflection of a conducting interface, the method advanced above for nonconducting media can be generalized readily. Starting again with the pair of equations (3.22), one must take into account the conducting properties of the medium. Replacing the matrix elements P_{mq} and Q_{mq} by the convolutions D_{mq} and S_{mq} containing the presentations of nonstationary fields (2.120)–(2.121), we obtain instead of eqn (3.48),

$$M_{mq}(\alpha) \to H_{mq}(\beta) = \frac{1}{t_0}\int_0^\infty L_m\left(\frac{t}{t_0}\right) e^{-\tau}[I_{q-1}(\tau) + I_{q+1}(\tau) - 2I_q(\tau)]\, dt, \tag{3.56}$$

$$N_{mq}(\alpha) \to S_{mq}(\beta) = \frac{1}{t_0}\int_0^\infty L_m\left(\frac{t}{t_0}\right) e^{-\tau}[I_{q-1}(\tau) - I_{q+1}(\tau)]\, dt, \tag{3.57}$$

$$\tau = tT_\sigma^{-1}, \quad \beta = t_0 T_\sigma^{-1}. \tag{3.58}$$

The time scale of the conducting material T_σ is defined in eqn (2.111). Let us introduce, by analogy with eqn (3.49), the matrices $D_{mq}(\beta)$ and $K_{mq}(\beta)$:

$$\left.\begin{matrix} D_{mq}(\beta) \\ K_{mq}(\beta) \end{matrix}\right\} = \frac{1}{2}[S_{mq}(\beta) \mp H_{mq}(\beta)], \tag{3.59}$$

$$\left.\begin{aligned} D_{mq}(\beta) &= \int_0^\infty L_m\left(\frac{t}{t_0}\right) e^{-\tau}[I_q(\beta\tau) - I_{q+1}(\beta\tau)]\, dt, \\ K_{mq}(\beta) &= \int_0^\infty L_m\left(\frac{t}{t_0}\right) e^{-\tau}[I_{q-1}(\beta\tau) - I_q(\beta\tau)]\, dt. \end{aligned}\right\} \tag{3.60}$$

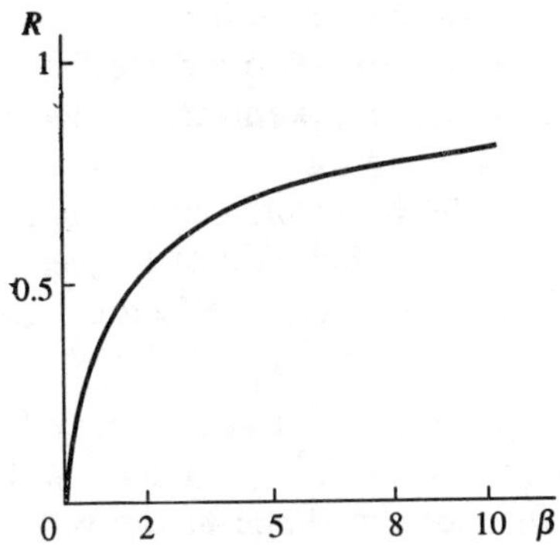

Fig. 3.6 Dependence of the coefficient of reflection of normally incident Laguerre waveforms R (eqn 3.65) from the interface of a conducting medium on the ratio of time scales $\beta = t_0 T_\sigma^{-1}$.

The equations (3.22) may be rewritten by means of the matrices $D_{mq}(\beta)$ and $K_{mq}(\beta)$ in the form

$$\sum_{q=1}^{\infty} a_q K_{mq}(\beta) = -2, \quad \sum_{q=1}^{\infty} a_q D_{mq}(\beta) = 2R. \tag{3.61}$$

Correlation between the matrix elements $K_{mq}(\beta)$ and $D_{mq}(\beta)$ may be obtained through the integral (Ryzhik and Gradstein 1980)

$$\int_0^{\infty} e^{-px} I_{q+1}(\beta x)\, dx = \beta B \int_0^{\infty} e^{-px} I_q(\beta x)\, dx, \tag{3.62}$$

$$B = [p + \sqrt{(p^2 - \beta^2)}]^{-1}. \tag{3.63}$$

Use of (eqn 3.62) results in the correlation

$$D_{0q}(\beta) = \beta B K_{0q}(\beta). \tag{3.64}$$

Combination of eqns (3.61) and (3.64) yields the analytical time-domain formula for the reflection coefficient R. Keeping in mind the value $p = 0.5 + \beta$, we obtain

$$R = \frac{2\beta}{1 + 2\beta + \sqrt{(1 + 4\beta)}}, \quad \beta = \frac{t_0}{T_\sigma}. \tag{3.65}$$

The graph $R(\beta)$ is depicted in Fig. 3.6.

Comparing the reflection of Laguerre pulses from dielectric and conducting interfaces, one can point out the following salient features of these processes:

1. Interactions of Laguerre pulses with the above-mentioned media are determined by the equality of reflection coefficients for all these pulses L_m, irrespective of number m, parameters α (eqn 3.55) or β (eqn 3.65) being equal. Therefore the reflection coefficients R (eqns 3.55 and 3.65) are not labelled with the index m. Unlike the Hermitian pulses (Fig. 3.4), an arbitrary linear combination of Laguerre waveforms does not possess any shape distortions due to reflection from plasma-like or conductive media.
2. The proportion of the energy of Laguerre waveforms projected into a reflecting material, $W = 1 - R^2$, parameter β being given, remains constant, independent of the shape of the waveform.
3. Introducing the time-domain analogy of reflection index for Laguerre pulses n_L via the Fresnel formulae (3.15),

$$n_{\mathrm{L}} = \frac{1 - R}{1 + R}, \tag{3.66}$$

one can see that, in view of eqns (3.55) and (3.65), the value n_{L} obeys the inequality $n_{\mathrm{L}} > 1$. This condition is shown in Section 3.4 to provide some general conclusions regarding the oblique incidence of such pulses on a reflecting interface.

3.4 Polarization phenomena in time-domain transient optics

Polarization effects in reflection–refraction problems become apparent in the case of oblique incidence of fields on the reflecting interface. Owing to the transverse polarization of a plane EM field in free space, such oblique incidence is associated with the existence of some components of this field which are orthogonal to the interface (Fig. 3.7). The magnetic (electric) normal component relates to S-(P)-polarization of the field. The reflection coefficients for both S- and P-polarized sinusoidal waves are given by the classical Fresnel formulae. When the interface separates free space and a continuous medium with refractive index n, these formulae can be written as

$$\left.\begin{aligned} R_{\mathrm{S}} &= \frac{\cos\gamma - \sqrt{(n^2 - \sin^2\gamma)}}{\cos\gamma + \sqrt{(n^2 - \sin^2\gamma)}}, \\ \mathrm{R}_{\mathrm{P}} &= \frac{n^2\cos\gamma - \sqrt{(n^2 - \sin^2\gamma)}}{n^2\cos\gamma + \sqrt{(n^2 - \sin^2\gamma)}}, \end{aligned}\right\} \tag{3.67}$$

where γ is the angle of incidence. The expressions (3.67) illustrate the polarization-dependent difference of reflection coefficients R_{S} and R_{P}, the other parameters of the problem (n, γ) being fixed. Notice that the

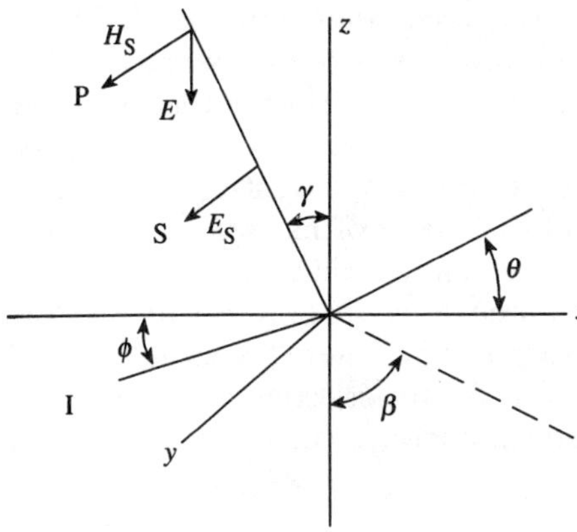

Fig. 3.7 Geometry of oblique incidence of S- and P-polarized waveforms on a reflecting interface I; γ and β are the angles of incidence and refraction respectively; the S(P)-polarized states are characterized by the $E_s(H_s)$ components orthogonal to the plane of incidence (x, z) and $H_{\perp S}(E_{\perp P})$ components normal to the interface I (x, y). The components $H_S(E_P)$ are parallel to the plane (x, z). Mixed polarization is characterized by the angles φ and θ between the component of electric field E parallel to the interface I and the x-axis in the incident and reflected waveforms respectively.

reflection coefficient for P-polarization fields is defined usually in terms of the ratio of the magnetic components of the reflected and incident fields, $H_{ref} = R_P H_0$ (Landau and Lifshitz 1984). Therefore it appears from eqns (3.67) that when $\gamma \to 0$,

$$R_P = \frac{n-1}{n+1}. \tag{3.68}$$

On the other hand, the reflection of S-polarized fields ($E_{ref} = R_S E_0$) in the limit $\gamma \to 0$ is described by eqn (3.15).

The generalization of these results for the reflection of polarized non-sinusoidal transients is based on the standard boundary conditions involving the continuity of the components of transient fields, both parallel and orthogonal to the reflecting interface. Therefore the boundary conditions become, in this case at least, three equations, unlike the pair of equations (3.22) applicable to normal incidence. The discussion of the reflection of both S-polarized (Subsection 3.4.1) and P-polarized (Subsection 3.4.2) waveforms reveals the essential influence of the shapes of incident waveforms on the polarization structure of the reflected signals.

These phenomena may hold potential for processing of picosecond and femtosecond transient signals in real time. Shape-dependent effects in reflection of single-cycle waveforms from plasma-like media are very pronounced when the eigenfrequency of the free carriers Ω_e and the pulse time scale t_0 obey the condition $\Omega_e t_0 \approx 1$–10. Reflectors produced from certain semiconductors with free carriers (e.g. InSb or GaAs) or ionic solids (e.g. TeO_2) and characterized by eigenfrequencies in the terahertz range open up possibilities for tuning of picosecond single-cycle pulses. Certain metal films (gold or aluminium oxide, $\Omega_e \approx 10^{15}\,\text{rad s}^{-1}$) may be of interest for shape-dependent processing of femtosecond waveforms. The difference in reflectlon coefficients R for these transients ($R^2 \ll 1$ or $R^2 \approx 1$) provides the physical basis for the selection of signals emerging from optical compressors for ultrafast spectroscopy and all-optical switching.

3.4.1 Reflection of S-polarized nonsinusoidal signals

To examine the boundary conditions for oblique incidence of EM waveforms, we should introduce the refraction angle β. Consideration of S-polarized waveforms, the continuity of the E_y, H_x and H_z components (Fig. 3.7) being taken into account, yields a set of equations determining the reflection coefficients R_S, transmission coefficients T_1, T_2, and refraction angle β:

$$1 + R_S = T_1, \tag{3.69}$$

$$(1 - R_S)\cos\gamma = T_2 \cos\beta, \tag{3.70}$$

$$(1 + R_S)\sin\gamma = T_2 \sin\beta. \tag{3.71}$$

For normal incidence ($\gamma = 1$), eqn (3.71) disappears, while eqns (3.69)–(3.70) become the system (3.11)–(3.12). The boundary conditions (3.69)–(3.71) are used below for analysis of the reflection of Laguerre and Hermitian waveforms from dispersive transparent media.

In view of the shape-independent reflection of Laguerre waveforms (see Subsection 3.3.1), it is worth beginning the consideration with this simpler case. Combining eqns (3.69) and (3.71), we obtain the analogy of Snell's law of refraction:

$$\sin\gamma = \frac{T_2}{T_1}\sin\beta. \tag{3.72}$$

Manipulating with eqns (3.69) and (3.70) and making use of eqn (3.72), one can find the analogy of the Fresnel formula (3.67):

$$R_S = \frac{\cos\gamma - \sqrt{(T_2^2 T_1^{-2} - \sin^2\gamma)}}{\cos\gamma + \sqrt{(T_2^2 T_1^{-2} - \sin^2\gamma)}}. \tag{3.73}$$

Thus the above-mentioned resemblance between the ratio $T_2 T_1^{-1}$ and

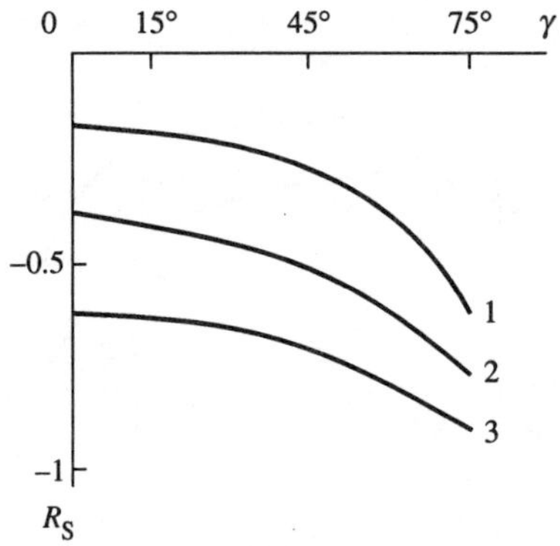

Fig. 3.8 Dependence of coefficient of reflection of S-polarized Laguerre waveforms from a dielectric interface R_S on the angle of incidence γ. Curves 1, 2, and 3 relate to the values of parameter $\Omega_e t_0 = 0.5$, 1, and 2 respectively.

refractive index (eqn 3.28) is displayed again: replacement of $T_2 T_1^{-1}$ by n results in coincidence of eqns (3.72) and (3.73) with the standard Snell and Fresnel formulae. Substitution of eqn (3.55) in eqn (3.27) yields the dependence of the ratio $T_2 T_1^{-1}$ on the dimensionless parameter $\alpha = \Omega_e t_0$ for plasma-like media:

$$\frac{T_2(\alpha)}{T_1(\alpha)} = \frac{\{1 + \sqrt{[1 + 4(\Omega_e t_0)^2]}\}^2 + 4(\Omega_e t_0)^2}{\{1 + \sqrt{[1 + 4(\Omega_e t_0)^2]}\}^2 - 4(\Omega_e t_0)^2}. \tag{3.74}$$

Finally, the substitution of eqn (3.74) in eqn (3.73) results in the time-domain reflection coefficient for Laguerre waveforms $R_S(\gamma, \alpha)$. Since $T_2 > T_1$, one can conclude that $R_S < 1$; this means that total internal reflection does not arise in the problem under discussion. The angular dependence $R_S(\gamma, \alpha)$ for these signals is shown in Fig. 3.8.

These general results are valid for Laguerre waveforms when the reflection coefficients can be determined separately from the transmission coefficients $T_{1,2}$. Although the reflection of Hermitian and Laguerre waveforms is examined using the same boundary conditions, the procedure for the analysis of Hermitian waveforms proves to be more complicated, since their reflection and transmission coefficients must be determined simultaneously. Equations (3.25)–(3.26) exemplify such an analysis for normal incidence; oblique incidence of Hermitian waveforms is examined in Subsection 3.5.1. The influence of the shapes of these waveforms on their reflection coefficients is illustrated in Fig. 3.9.

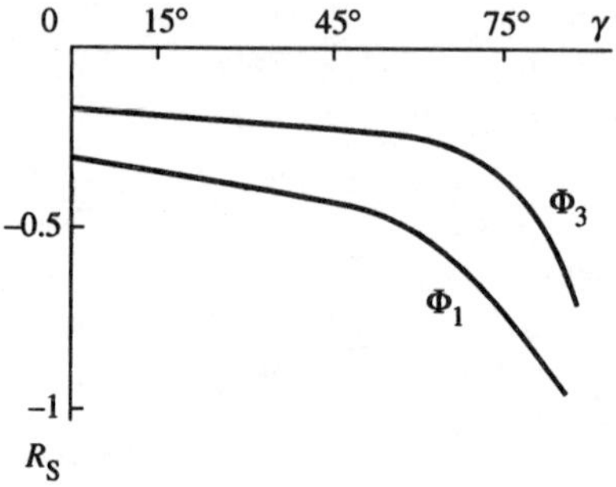

Fig. 3.9 Shape-dependent reflection of different S-polarized Hermitian waveforms Φ_1 and Φ_3 from a dielectric with free carriers ($\Omega_e t_0 = 2$). The reflection coefficient R_s is plotted vs. the angle of incidence γ.

3.4.2 Brewster effect for P-polarized single-cycle pulses

The continuity conditions for a P-polarized field incident on a dielectric interface may be written by analogy with the system (3.69)–(3.71):

$$1 + R_P = T_2, \tag{3.75}$$

$$(1 - R_P)\cos\gamma = T_1 \cos\beta, \tag{3.76}$$

$$(1 + R_P)\sin\gamma = d\sin\beta. \tag{3.77}$$

Equation (3.77) is associated with the continuity of the normal component of electric displacement d. Because of this function, which did not arise in the analysis of S-polarized fields (Subsectlon 3.4.1), a new approach to the consideration of P-polarized waveforms is required.

The electric displacement d was presented in eqn (2.56) as a sum of nonseparable harmonics d_q. Introducing an auxiliary function

$$K_q(\tau) = I_{q-1}(\tau) + I_{q+1}(\tau), \tag{3.78}$$

we can write explicit expressions for even ($q = 2n$) and odd ($q = 2n - 1$) harmonics at the boundary interface, $z = 0$:

$$d_{2n}(\tau) = -\frac{1}{2}\left[K_{2n}(\tau) + 2\sum_{p=1}^{\infty} K_{2n+4p}(\tau)\right] \tag{3.79}$$

$$d_{2n-1}(\tau) = -\frac{1}{2}\left[K_{2n-1}(\tau) + 2\sum_{p=1}^{n-1} K_{2p-1}(\tau)\right], \tag{3.80}$$

$$n = 1, 2, 3, \ldots$$

Thus, owing to the difference in the integrals from even and odd Bessel functions, the harmonics d_{2n-1} and d_{2n} contain either finite sums or infinite series respectively.

It is noteworthy that the function $K_q(\tau)$ is proportional to the qth harmonic of the magnetic component h_q (eqn 2.38): $K_q = 0.5 \cdot h_q$. Thus the harmonics d_q, eqns (3.79)–(3.80), prove to be expressed in terms of the sums of the magnetic harmonics h_q. This result permits one to analyse the system (3.75)–(3.77) by means of the pairs of matrix elements P_{mq}, Q_{mq} (eqn 3.21) or M_{mq}, N_{mq} (eqn 3.48) introduced above.

Let as consider for example the reflection of P-polarized Hermitian waveforms Φ_m, representing $d(\tau)$ (eqn 2.56) by analogy with $E(\tau)$ and $H(\tau)$ (eqn 3.18):

$$\left.\begin{aligned} d(\tau)\,|_{\eta=0} = \sum_m \Phi_m(\tau)\cdot D_m, \quad D_m = \sum_{q=1}^{\infty} D_{mq} a_q, \\ D_{mq}(\alpha) = \frac{1}{2t_0}\int_{-\infty}^{\infty} \Phi_m\left(\frac{t}{t_0}\right) d_q(\Omega_e t)\,\mathrm{d}t. \end{aligned}\right\} \tag{3.81}$$

Substitution of eqn (3.79) or eqn (3.80) yields the different matrix elements D_{mq} for the even and odd harmonics d_q. Thus, considering for simplicity the odd ones, we obtain from eqn (3.80)

$$D_{mq} = -\frac{1}{2}\left[Q_{mq} + 2\sum_{p=1}^{n-1} Q_{m,2p-1}\right]. \tag{3.82}$$

The system (3.75)–(3.77) describing the reflectlon of the mth waveform may be written as

$$\left.\begin{aligned} 1 + R_m = T_{2m}, \quad (1 - R_m)\cos\gamma = T_{1m}\cos\beta_m, \\ (1 + R_m)\sin\gamma = D_m \sin\beta_m. \end{aligned}\right\} \tag{3.83}$$

Here T_{1m}, T_{2m}, and D_m are the linear combinations of unknown coefficients a_q, eqns (3.19), (3.20), (3.81). These coefficients have to be determined from the system (3.83) together with the reflection coefficients R_m and refraction angles β_m. The truncation of the infinite system (3.83) is performed by analogy with eqns (3.25)–(3.26). Thus, considering the reflection of the Φ_2 waveform (Fig. 1.9) we can write the simplest truncated system by means of matrix elements P_{mq} and Q_{mq} for $m = 2$ and $m = 0$:

$$\left.\begin{aligned} 2(1 + R_2) = \sum_{n=1}^{4} Q_{2,2n-1} a_{2n-1}, \\ 2(1 - R_2)\cos\gamma = \cos\beta_2 \sum_{n=1}^{4} P_{2,2n-1} a_{2n-1}, \end{aligned}\right\} \tag{3.84}$$

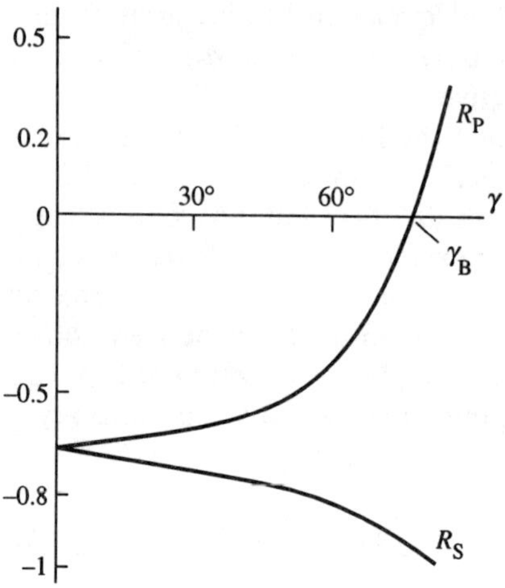

Fig. 3.10 Coefficient of reflection of P- and S-polarized Hermitian waveforms Φ_2 vs. angle of incidence γ ($\Omega_e t_0 = 2$). The angle $\gamma_B = 78°$ relates to the Brewster angle for the waveform Φ_2 ($R_P = 0$).

$$\left.\begin{aligned} 2(1+R_2)\sin\gamma = \sin\beta_2 \sum_{n=1}^{4} D_{2,2n-1}a_{2n-1}, \\ \sum_{n=1}^{4} Q_{0,2n-1}a_{2n-1} = 0, \quad \sum_{n=1}^{4} P_{0,2n-1}a_{2n-1} = 0, \quad \sum_{n=1}^{4} D_{0,2n-1}a_{2n-1} = 0. \end{aligned}\right\} \quad (3.85)$$

Equations (3.85) show that the waveform with $m = 0$ is absent in the incident field.

The system (3.84)–(3.85) contains six equations and six unknowns: the reflection coefficient R_2, the refraction angle β_2, and the coefficients a_1, a_3, a_5, a_7. The solution of this system depends on the value of the parameter $\alpha = \Omega_e t_0$ and the angle of incidence γ. To expand this system, values $m = 2$, 4, and so on can be taken into account.

The dependence of the reflection coefficient for a P-polarized Φ_2 waveform on the angle of incidence is shown in Fig. 3.10. An important feature of this graph is associated with the vanishing of reflection ($R_P = 0$) a some value of parameter γ. This is an analogy of the Brewster effect, well known in the optics of sinusoidal waves: tg $\gamma_B = n$ (eqn 1.189). However, the Brewster angle γ_B for nonsinusoidal waveforms depends on the shape of this waveform, the value of parameter $\alpha = \Omega_e t_0$ being fixed.

Mixed polarization of the incident waveform can result in rotation of the plane of polarization owing to a difference in reflection coefficients for the

S- and P-polarized components. Thus the component of the electric field of the incident field parallel to the reflecting interface may be characterized in the case of mixed polarization by an angle φ between this component and the x-axis. The component $E_{\parallel}$ in the reflected field is characterized by another angle θ (Fig. 3.7). Using the reflection coefficients for S- and P-polarizations, one can find the correlation between these angles:

$$\operatorname{tg}\theta = \frac{R_S}{R_P}\operatorname{tg}\varphi. \tag{3.86}$$

Owing to the dependence of coefficients R_S and R_P on the shape of the waveform, this rotation of the polarization plane of the reflected pulse also proves to be shape-dependent. This rotation as well as shape-dependent Brewster effect may be of interest for the selection of single-cycle pulses with respect to their envelopes.

3.5 Nonstationary refracted fields in dispersive media

Spatial–temporal structures of refracted EM fields excited by ultrashort transients in a dispersive medium are determined by the envelopes of the nonseparable harmonics e_q, h_q, and the relevant coefficients a_q, eqns (2.36)–(2.37). The dynamics of these structures is distinguished in principle from the well-known process of slowly developing dispersive distortions of long pulses containing many oscillations of sinusoidal fields (Jackson 1975). The characteristic spatial scale of accumulation of such distortions—the so-called 'dispersion length' L_ω (eqn 1.60)—is known to be much larger than the length of the pulse itself. In contrast, the rapid deformations of the nonseparable harmonics e_q, h_q, described in Chapter 2, provide the following features of dispersive spreading of short waveforms in a medium with plasma-like dispersion:

1. The accumulation of distortions results in total reshaping of the single-cycle pulse in such a medium. The envelope of the refracted signal may have nothing in common with the profile of the exciting waveform; thus the wings of an incident single-cycle pulse are transformed in the course of propagation to a sequence of quasiharmonic oscillations (eqn 3.34).
2. Unlike the dispersive evolution of a long narrow-band wave pulse, the envelopes of the magnetic and electric components of a refracted single-cycle waveform may become very different in the course of propagation.
3. The process of reshaping of the refracted waveform is very rapid: the distance of reshaping, being much shorter than the 'dispersive length'

L_ω (eqn 1.60), may become comparable with the length of the exciting pulse.

Although the analytical background to the consideration of refracted nonstationary fields based on continuity conditions was described above, some physically meaningful aspects of such problems are discussed below. It was pointed out above that the problems of reflection and refraction for Hermitian waveforms must be considered simultaneously, whereas the same problems for Laguerre waveforms may be examined separately. From the formal viewpoint, this means that the reflection and refraction coefficients R and T_1, T_2 for Hermitian (Laguerre) waveforms have to be determined together (separately).

3.5.1 Refraction of Hermitian waveforms

It was pointed out above that the nonzero matrix elements P_{mq} and Q_{mq} (eqn 3.21) contain the convolutions of Hermitian envelopes and Bessel functions of the same parity. An example of the use of this 'selection rule' for odd envelopes ($m = 2n - 1$) were given by eqns (3.25)–(3.26). Some interesting properties of refracted fields excited by symmetrical (even) envelopes Φ_m ($m = 2n$) are outlined below.

1. Let us consider the normal incidence of, for example, the Φ_2 waveform. The continuity conditions at the interface $\eta = 0$ may be written by analogy with eqn (3.25):

$$\left.\begin{aligned} 1 + R_2 &= P_{21}a_1 + P_{23}a_3 + P_{25}a_5 + \dots \\ 1 - R_2 &= Q_{21}a_1 + Q_{23}a_3 + Q_{25}a_5 + \dots \end{aligned}\right\} \tag{3.87}$$

The analogous equations containing the matrix elements P_{mq} and Q_{mq} with $m \neq 2$ have the left-hand side equal to zero (eqn 3.26). By finding the coefficients a_q from the relevant truncated system of algebraic equations, one can take notice of the coefficient a_1 determining the excitement of harmonics e_1 and h_1 of the refracted field:

$$\left.\begin{matrix} e_1 \\ h_1 \end{matrix}\right\} = a_1(\Psi_0 \mp \Psi_0). \tag{3.88}$$

Here Ψ_0 and Ψ_2 are the nonseparable harmonics (eqn 2.27).

Unlike any other harmonics wlth $q \neq 1$, the harmonics e_1 and h_1 preserve their peak amplitudes a_1 in the course of propagation in a lossless medium with plasma-like dispersion (Fig. 3.11). This property may be of interest for communication and remote-sensing purposes.

2. The time-domain generalization of this approach reveals the specific nature of the magneto-optics of refracted transients based on the birefringence of the refracting medium. Let us examine the normal incidence

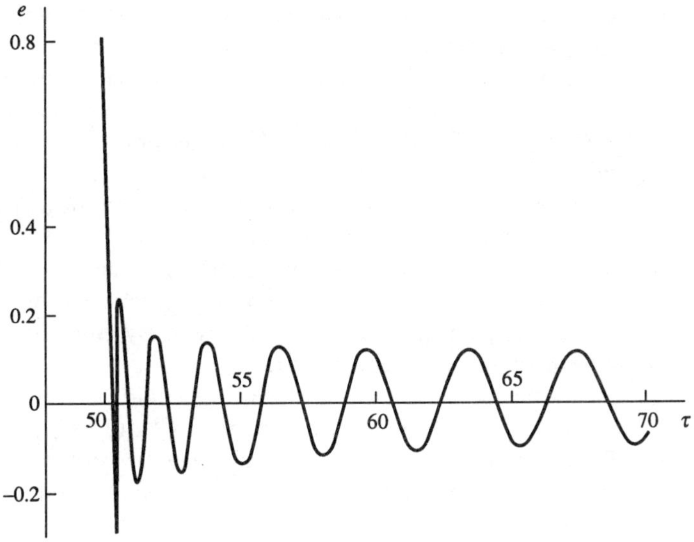

Fig. 3.11 Electric component of the refracted field excited by Hermitian waveform Φ_2 incident normally on the interface of a plasma-like medium ($\eta = 0$), at a position $\eta = 50$ within the medium ($\Omega_e t_0 = 2$). The normalized envelope of electric component e is plotted vs. the dimensionless local time $\tau_1 = \tau + \eta$.

of the waveforms Φ_m on the interface of a magnetoplasma layer, an external magnetic field H_0 being parallel to this interface. Designating the angle between the plane of polarization of the waveform and the field H_0 as δ, one can discuss the refraction of two waveforms with electric fields $E_1 = E_0 \cos\delta$ ($E_1 \| H_0$) and $E_2 = E_0 \sin\delta$ ($E_2 \perp H_0$). The properties of the refracted fields excited by these waveforms are known to be very different (see Subsection 2.4.1).

As to waveform E_1, the magnetic field has no influence on reflection and refraction of this polarization ($E_1 \| H_0$). Therefore the coefficients R_m and a_q are determined from the system (3.25) or (3.87) by means of matrix elements $P_{mq}(\alpha)$ and $Q_{mq}(\alpha)$. Unlike this ordinary polarized refracted field, the refraction of waveform E_2 excites the extraordinary polarized nonseparable harmonics (eqn 2.37) travelling across the magnetic field H_0; if this field is strong enough ($\omega_H^2 \gg \Omega_e^2$), the time scales of these harmonics are $t_\perp = \omega_H \Omega_e^{-2}$, $t_\perp t_1^{-1} = \omega_H \Omega_e^{-1} > 1$. Therefore the matrix elements determining the reflection and refraction of this extraordinary polarized waveform may be written as $P_{mq}(\alpha\Omega_e\omega_H^{-1})$ and $Q_{mq}(\alpha\Omega_e\omega_H^{-1})$. Substituting these matrix elements in eqn (3.25) or (3.87), one can use the standard procedure for determination of reflection and refraction coefficients for this waveform. Since the spatial and temporal scales are polarization-dependent, the structures of the ordinary and extraordinary polarized refracted fields at any given space–time point differ. Moreover, the frequencies of the quasi-harmonic oscillations, Ω_e and $\Omega_e^2\omega_H^{-1}$, which were shown to arise at the wings of each refracted field, also prove to be discernible.

3. When considering the refraction of obliquely incident P-polarized EM fields (Subsection 3.4.2), we presented the electric and magnetic components of the refracted fields in the form of eqn (2.37). The matrix elements P_{mq} and Q_{mq} determining the refraction coefficients T_{1m} and T_{2m} were given in this case by the simple formula (3.48). whereas the components of the electric displacement d_{2n} were expressed by complicated formulae containing infinite sums, eqn (3.79).

To avoid calculation of these infinite sums, one could present the non-separable harmonics e_q, h_q, and d_q in the forms (2.42), (2.43), and (2.55) respectively. The components of the electric and magnetic fields and electric displacement may be written in this case by means of the functions P_{mq} and Q_{mq} (eqn 3.48):

$$T_{1m} = \sum_{q=2}^{\infty} A_{mq} b_q, \quad T_{2m} = \sum_{q=2}^{\infty} B_{mq} b_q, \quad D_m = \sum_{q=2}^{\infty} C_{mq} d_q, \tag{3.89}$$

$$\left.\begin{aligned} A_{mq} &= \frac{1}{2}(P_{m,q-1} - P_{m,q+1}), \quad B_{mq} = \frac{1}{2}(Q_{m,q-1} + Q_{m,q+1}), \\ C_{mq} &= \frac{1}{2}(Q_{m,q-1} + Q_{m,q+1}). \end{aligned}\right\} \tag{3.90}$$

Although the expressions for the matrix elements A_{mq} and B_{mq} (eqn 3.90) are slightly more complicated than the related expressions P_{mq} and Q_{mq}, the matrix element C_{mq} (eqn 3.90) does not contain infinite sums. Hence this representation of an EM field, introduced in Subsections 2.2.2 and 2.3.1, may be used to simplify the analysis of reflection–refraction problems.

Although these refraction phenomena are power-independent, the preservation of narrow peaks of e_1 and h_1 (eqn 3.88) may be reminiscent of nonlinear soliton propagation in a dispersive medium. However, the soliton envelope contains many oscillations, and a fixed correlation between its peak amplitude and halfwidth is required. The propagation of harmonics (eqn 3.88) into the depth of a plasma-like medium is accompanied by an increase in contrast between the preserved peaks and spreading wings.

3.5.2 Excitation of nonseparable fields by Laguerre transients

Since the reflection of Laguerre pulses from plasma-like materials has been shown to be shape-independent (Section 3.3), there is no influence of the envelope of incident Laguerre waveforms on their refraction coefficients

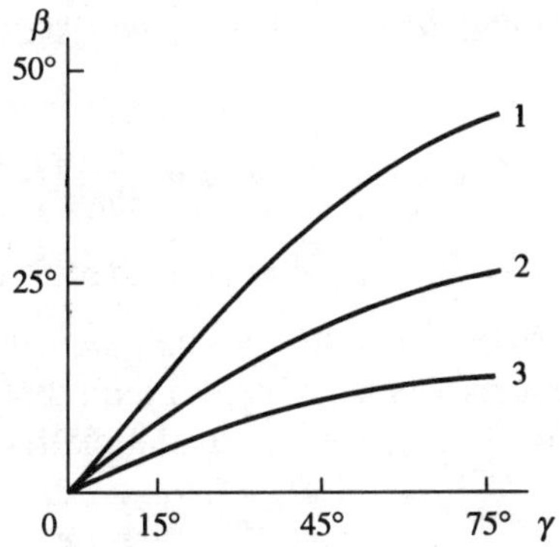

Fig. 3.12 Angle of refraction of Laguerre pulses β in dielectrics with free carriers, plotted against the angle of incidence γ. Curves 1, 2, and 3 relate to the values of parameter $\Omega_e t_0 = 0.5$, 1, and 2 respectively.

T_1 and T_2. The proportion of the energy of the incident waveform projected into the medium depends only on the value of parameter $\alpha = \Omega_e t_0$, the angle of incidence γ being glven:

$$W = 1 - R^2 = T_1(\alpha) \cdot T_2(\alpha). \tag{3.91}$$

The refraction angle β also proves to be independent of the envelope of the incident waveform. The values $\beta = \beta(\Omega_e t_0)$ determined from eqn (3.72) are shown in Fig. 3.12. Hence valuable information concerning the refracted field may be obtained without any consideration of the coefficients a_q describing the excited nonseparable fields.

The coefficients a_q characterizing the refracted waveform may be found from a system of algebraic equations, similar to the system (3.25)–(3.26), associated with Hermitian waveforms. Considering for instance the reflection of the Laguerre waveform L_1 (eqn 3.5), described by reflection coefficient R, one can write the system determining the coefficients a_q in the form

$$\sum_{q=1}^{\infty} E_{mq} a_q = (1 + R)\delta_{m1}, \tag{3.92}$$

$$\sum_{q=1}^{\infty} G_{mq} a_q = (1 - R)\delta_{m1}. \tag{3.93}$$

Here δ_{m1} is the Kronecker delta, and the matrix elements $E_{mq}(\alpha)$ and $G_{mq}(\alpha)$ are expressed in terms of X_{mq} and Y_{mq} (eqn 3.50):

$$\left.\begin{matrix} E_{mq}(\alpha) \\ G_{mq}(\alpha) \end{matrix}\right\} = \frac{1}{2}\left[Y_{mq}(\alpha) \mp X_{mq}(\alpha)\right]. \tag{3.94}$$

These matrix elements may be calculated by means of the integral (Ryzhik and Gradstein 1980)

$$\int_0^\infty e^{-px} x^n I_q(\alpha x)\,dx = (-1)^n \alpha^q \frac{\partial^n}{\partial p^n}(BA^q), \tag{3.95}$$

$$B = (p^2 - \alpha^2)^{-1/2}, \quad A = [p + \sqrt{(p^2 - \alpha^2)}]^{-1}. \tag{3.96}$$

Using these formulae with the value $p = 0.5$, we can find the analytical expressions for the matrix elements $Y_{0q}(\alpha)$ (eqn 3.50); finally, keeping in mind the correlation (3.54), we obtain the matrix elements (3.94) in the form

$$\left.\begin{matrix} E_{0q}(\alpha) \\ G_{0q}(\alpha) \end{matrix}\right\} = \frac{1}{2}(1 \mp \alpha^2 A^2)\, Y_{0q}(\alpha). \tag{3.97}$$

Substitution of eqn (3.97) in the infinite system (3.92) and solution of the truncated system yields the values of the coefficients a_q in the nonseparable representation of the refracted field.

Thus the refraction of both Hermitian and Laguerre waveforms in a medium with plasma-like dispersion excites nonstationary, rapidly deforming EM fields. Similar fields were shown to exist in waveguide systems, characterized by the same dispersion law (Subsection 2.6.1). This analogy is considered below for the experimental modelling of refraction of ultrashort single-cycle waveforms.

3.6 Modelling of dynamics of nonsinusoidal waveforms in dispersive systems

The time-domain description of the interaction of transients with dispersive systems has traditionally been associated with computer simulations. The explicit nonstationary solutions obtained above represent the first examples of straightforward time-domain analysis without the use of inverse Fourier transforms. To be convinced of the validity of these results, one would need physical modelling. However, the large spatial scales of nanosecond pulses hinder laboratory observations of their interaction with the bulk of a plasma-like medium. As to subfemtosecond waveforms emerging from an optical compressor, relevant experiments are hampered by the problem of controlled shaping of such waveforms. Flexible tuning of transient shapes and reasonable scales as well as programmable variation of the characteristic dispersion and attenuation time scales of the modelling system are therefore required for laboratory modelling of these transient EM processes. Two types of such system, corresponding to lossless and conducting media, are considered below.

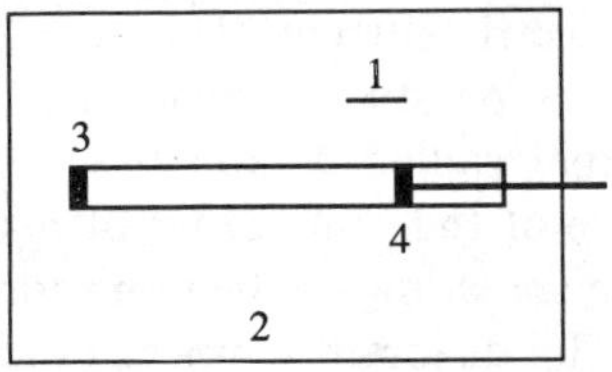

Fig. 3.13 Scheme of setup for cryoacoustic modelling of shape-dependent propagation of nonsinusoidal waveforms in a dispersive medium. The second-sound waveguide 1 (a capillary filled with superfluid helium-II) is immersed in a cryostat 2; 3 is a thin-film emitter, and 4 a mobile bolometer recording the cryoacoustic waveforms.

The excitation of nonseparable harmonics in weakly attenuating systems with plasma-like and waveguide-like dispersion can be simulated by means of the superfluidity effect, discovered by Kapitza as far back as 1938. One of the manifestations of this effect is associated with so-called second sound waves in superfluid helium-II. These waves represent temperature oscillations propagating in the superfluid helium with extremely low velocity, $u \approx 20\,\mathrm{m\,s^{-1}}$, the helium temperature being as low as $t < 2.19\,\mathrm{K}$. Such cryoacoustic oscillations are known to be excited effectively by a source with varying temperature (Putterman 1974). This velocity, which is 10^7 times smaller than the velocity of light in free space, permits one to avoid the aforesaid problems inherent in the use of large-scale nanosecond EM pulses. The tunable temperature waveforms in cryoacoustic systems can be shaped much more easily than the ultrashort nonsinusoidal transients in optoelectronic devices. At least, the capillaries filled with superfluid helium act as waveguides for second-sound signals, providing dispersion of these signals. Such systems, consisting of an emitter of temperature waveforms, a waveguide dispersive element, and a sensitive bolometer receiving these temperature signals, may be of interest for modelling of electromagnetic processes in dispersive media (Shvartsburg 1989).

Simulation of EM pulse phenomena in absorbing media can be performed through excitation of nonsinusoidal current and voltage waveforms in a transmission line.

3.6.1 Dispersive reshaping of cryoacoustic pulses in capillary waveguides

Consider the cryoacoustic modelling system shown in Fig. 3.13. A capillary tube 1, filled with superfluid helium-II, is located inside a cryostat 2. A thin metallic film 3, heated by a pulse electric current, emits solitary temperature waveforms with time scales $t_0 \sim 10^{-4}$–10^{-5} s. The time of this film heating ($\tau_T \leqslant 10^{-6}$ s) permits the temperature waveform to be shaped by control of the envelope of the electric current. The nonsinusoidal pulses emitted, containing one or a few temperature oscillations, travel along the capillary waveguide, whose internal radius ρ_0 and the initial length of the pulse $1 = ut_0 \approx 0.1$–0.01 cm are of the same order of magnitude. This condition provides rapid dispersive spreading of the nonstationary temperature fields excited by the emitter. The distorted temperature profiles are recorded by bolometer 4. By moving the bolometer, one can vary the path of dispersive evolution. The signal from the bolometer represents the result of this simulation.

Representing the temperature perturbation in the capillary in terms of its maximum value $(\Delta t)_m$ and the dimensionless function $f\,[\Delta t = f \cdot (\Delta t)_{\mathrm{m}}]$, we can find the function f from the wave equation governing thc second-sound propagation:

$$\frac{\partial^2 f}{\partial \rho^2} + \frac{1}{\rho}\frac{\partial f}{\partial \rho} + f + \frac{1}{\rho^2}\frac{\partial^2 f}{\partial \varphi^2} + \frac{\partial^2 f}{\partial z^2} - \frac{1}{u^2}\frac{\partial^2 f}{\partial t^2} = 0. \tag{3.98}$$

Here ρ, φ, and z are the cylindrical coordinates associated with the capillary tube, and u is the second-sound velocity. Restricting ourselves to the axisymmetrical temperature distributions and generalizing the scheme of plane waveguide analysis developed above in Subsection 2.6.1, we can present the solution of eqn (3.98) in a form similar to eqn (2.149):

$$f(\rho, z, t) = \sum_{n=0}^{\infty} \sum_{m=1}^{\infty} \sum_{q=1}^{\infty} a_{nmq} \cdot J_n(k_{nm}\rho) \cdot \Psi_q(\tau_{nm}, \eta_{nm}). \tag{3.99}$$

Here J_n is the Bessel function and the eigenvalues k_{nm} are determined from the boundary condition at the tube's surface,

$$\left.\frac{\partial f}{\partial \rho}\right|_{\rho=\rho_0} = 0, \quad k_{nm} = \frac{\nu_{nm}}{\rho_0}, \tag{3.100}$$

where ν_{nm} is the mth root of the nth order Bessel function. The functions $\Psi_q(\tau_{nm}, \eta_{nm})$ are the nonseparable solutions of the Klein–Gordon equation (2.27), and the normalized variables τ_{nm} and η_{nm} are

$$\eta_{nm} = z \cdot k_{nm}, \quad \tau_{nm} = u \cdot t \cdot k_{nm}. \tag{3.101}$$

The spatial–temporal structure of the field (3.99), depending on the three indices n, m, and q, is obviously more complex than the structure (2.149) relating to a plane waveguide. To simplify the problem, let us consider the pulse axisymmetrical heating current travelling from the centre to the periphery of the circular flow emitter 3 (Fig. 3.13). This current will provide a radially monotonic distribution of the nonstationary heat flow $W(t)$; greatest at the centre, this flow decreases towards the periphery. Therefore the main contribution to this flow is provided by the zeroth-order Bessel function ($n = 0$, $\nu_{01} = 2.4$, $\nu_{02} = 5.52$, $\nu_{03} = 8.65$).

The dimensionless value $\eta_{01} = 100$, corresponding to the complete dispersive distortion of the travelling nonseparable harmonics, determines the distance of propagation $z \sim 1$ cm; the distances relating to η_{0m} $(m > 1)$ are shorter by a factor of $\nu_{01} \cdot \nu_{0n}^{-1}$. Since the excitation of higter modes $(n > 0)$ is weakened in the said geometry, and the higher harmonics $(m > 1,\ q > 1)$ spread more rapidly, the structure of the nonseparable temperature field at a distance of ~ 1 cm wlll be determined mainly by the fundamental mode $J_0(k_{01}\rho)$ and the first harmonics Ψ_q (eqn 3.99).

The attenuation of these second-sound fields with time scales of $\sim 10^{-4} - 10^{-5}$ s, propagating in narrow waveguides ($\rho_0 \approx 0.05$ cm) depends basically on the surface losses (Putterman 1974). Considering the helium temperature, $t = 1.87$ K (this value is chosen in order to eliminate the nonlinear dependence of the second-sound velocity u on its amplitude), one can find the attenuation length to be $L_a \approx 30$–50 cm. Therefore the absorption of these fields, the path of propagation $l \approx 1$–2 cm being much shorter than the attenuation length L_∞, may be neglected.

Registration of nonstationary temperature fields with the peak amplitudes $(\Delta t)_m \approx 1 - 2 \times 10^{-2}$ K by a sensitive bolometer makes it possible to simulate the rapid evolution of refracted waveforms in dispersive systems, visualizing for example the following phenomena discussed above:

(1) shock generation of cutoff frequencies $\omega_{nm} = u \cdot k_{nm}$ inside the waveguide;

(2) preservation of a narrow peak on the background of a spreading field envelope;

(3) accelerated spreading of the higher nonseparable harmonics Ψ_q.

The salient features of this cryoacoustical modelling system are:

(1) use of microsecond cryoacoustic pulses instead of picosecond and subfemtosecond optical waveforms;

(2) simplicity and flexibility of shaping of the pulse heat flow $W(t)$ projected into the capillary waveguide;

(3) use of high-resolution receivers of the second-sound waveforms, recording their temporal envelopes.

Moreover, this system can be used directly for modelling the propagation of ultrashort transients in cylindrical waveguides and weakly attenuating transmission lines.

3.6.2 Dynamics of Laguerre transients in transmission lines

Reflection and refraction of ultrashort transients at the interface of a conducting medium can be modelled by means of a transmission line. The spatial–temporal distributions of the nonsinusoidal current i and voltage u

in this line were obtained above, eqns (2.137)–(2.140). To find the reflection and transmission coefficients for the Laguerre transients applied to the end of this line, one can use the continuity conditions for current and voltage at the end $z = 0$. Let us suppose for simplicity that the voltage and current transients represented by Laguerre envelopes are incident on this end $z = 0$ from some distortion-free transmission line.

Since the transmission line is characterized by two different time scales t_1 and t_2 (eqn 2.130) and different expressions for the voltage (eqns 2.139 and 2.140), which are valid in the cases $t_1 < t_2$ and $t_2 < t_1$ respectively, these cases must be analysed separately. Let us consider first eqn (2.140). Keeping in mind the equality of the reflection coefficients R for all the Laguerre pulses (see Subsection 3.2.2), we can write the continuity conditions using the designations (3.56)–(3.57):

$$-2(1-R) = \sum_{q=1}^{\infty} S_{mq} a_q, \tag{3.102}$$

$$-2(1+R) = U \sum_{q=1}^{\infty} H_{mq} a_q, \quad U = \frac{1}{2}\left|\frac{t_1}{t_2} - 1\right|. \tag{3.103}$$

Generalizing the analysis developed above, eqn (3.60)–(3.64), we obtain the formula for the reflection coefficient:

$$R|_{t_1 > t_2} = \frac{1 + pB - U(1 - pB)}{1 + pB + U(1 - pB)}, \tag{3.104}$$

where

$$p = \frac{1}{2} + \alpha\beta. \tag{3.105}$$

Parameter α, depending on the time scales t_1 and t_2, is defined in eqn (2.136); parameter β is the ratio of the time scales of the Laguerre pulse t_0 and the transmission line $T_0 = t_1 U^{-1}$ eqn (2.130):

$$\beta = t_0 T_0^{-1}. \tag{3.106}$$

Parameter B is given by eqn (3.63). The analysis in the opposite case, $t_1 < t_2$, is performed analogously.

The dependence of the reflection coefficients R on the parameter β, eqn (3.106), for different ratios $t_1 t_2^{-1}$ is depicted in Fig. 3.14. The refracted pulses can be measured at different positions along the line; the envelopes of current i and voltage u simulate the nonstationary magnetic and electric fields in the conducting medium by varying the inductance, resistance, and capacity of transmission line, i.e. by varying the time scales t_1 and t_2 (eqn 2.130), one can use this modelling system for optimization of the characteristics of an EM field, penetrating a conducting medium.

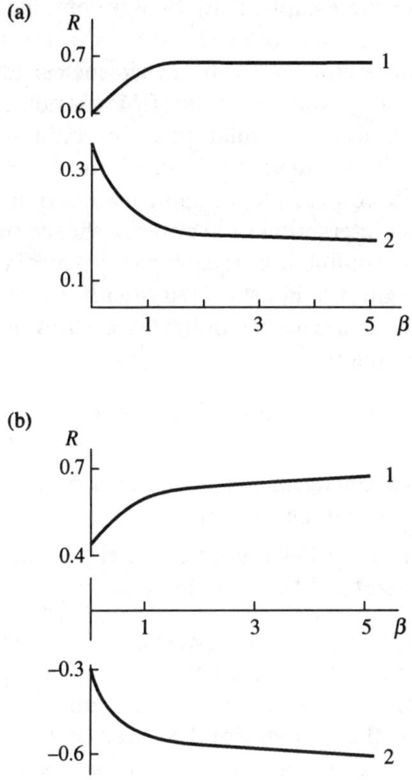

Fig. 3.14 Reflection of Laguerre waveforms from the end of a transmission line. The reflection coefficient R is plotted vs. the normalized time scale $\beta = t_0 T_0^{-1}$. (a) Curves 1 and 2 relate to values of parameters of the line $t_2 t_1^{-1} = 2$ and $t_2 t_1^{-1} = 0.5$ respectively. (b) Curves 1 and 2 correspond to the ratios $t_2 t_1^{-1} = 5$ and $t_2 t_1^{-1} = 0.2$ respectively. The definitions of characteristic times t_1 and t_2 are given by eqn (2.130).

3.7 Conclusion: Fourier and non-Fourier optics: complementary tools in investigation of ultrashort waveforms

1. At first sight, Fourier and non-Fourier optics could be associated directly with frequency-domain and time-domain descriptions of electromagnetic phenomena. In accordance with this viewpoint, one should attribute any result formulated without the use of frequency-related concepts to non-Fourier optics, and vice versa. However, the difference between these approaches is sometimes not so obvious. Thus the dependence of the absorption of solitary waveforms in Debye media on their time scales and shapes is described in Subsection 3.2.2. (Fig. 3.5) beyond the scope of Fourier

expansions, although these explicit results were obtained by integration of the related convolutions of Fourier transforms in the frequency domain. Very rarely however does this integration yield such exact analytical results; the time-domain descriptions of pulse EM phenomena were obtained basically either by numerical calculation of inverse Fourier transforms or by finite-difference time-domain solution of the Maxwell equations (see Subsection 1.8.1). In contrast, a straightforward time-domain analytical approach to the rapid interaction of arbitrarily shaped nonsinusoidal waveforms with dispersive, conducting, and magnetized media is advanced above.

Providing new insight into nonstationary fields, the time-domain electrodynamics of continuous media has indicated a series of new transient EM phenomena, for example:

(a) shape-dependent discrimination of single-cycle waveforms in reflection–refraction processes;
(b) existence of invariant elements in the structure of rapidly spreading refracted fields in guiding systems;
(c) difference in the rates of evolution of the electric and magnetic components of nonseparable EM fields in conductors.

From the formal viewpoint, these results are encoded in the relevant Fourier representations in the frequency domain. However, the extraction of such explicit results from these integral presentations is a target hard to attain.

2. Together with the analysis of EM transients in dispersive media, non-Fourier electrodynamics may also become a powerful tool for the consideration of reflection and scattering phenomena in free space. These problems are now becoming real, because of the design of curvilinear reflectors forming transient fields with some special properties in nanosecond radiophysics and picosecond optics. The traditional approaches to these problems are associated with complicated computer simulations, using for example the finite-difference time-domain method for scattering of spherical waves (Rao and Wilton 1991) or special computer codes for numerical analysis of pulse responses of certain curvilinear surfaces (Sun and Rusch 1994). In contrast, exact analytical solutions in time-domain pulse electrodynamics may be obtained by means of the special representation of an EM field given by Bateman (1955).

Following this representation, let us introduce the vector $\boldsymbol{M}$ as a linear combination of the electrlc $\boldsymbol{E}$ and magnetic $\boldsymbol{H}$ components of an EM field in a free space:

$$\boldsymbol{M} = \boldsymbol{H} + \mathrm{i}\boldsymbol{E}. \tag{3.107}$$

The equation governing vector $\boldsymbol{M}$ can be derived from the Maxwell equations

$$\nabla \times M = -\frac{\mathrm{i}}{c}\frac{\partial M}{\partial t}. \tag{3.108}$$

Further, representing vector M by scalar fields U and Ψ in the form

$$M = \nabla U \times \nabla \Psi, \tag{3.109}$$

and substituting eqn (3.109) in eqn (3.108), we obtain an equation describing these scalar fields:

$$\nabla U \times \nabla \Psi = -\frac{\mathrm{i}}{c}\left(\frac{\partial U}{\partial t}\nabla \Psi - \frac{\partial \Psi}{\partial t}\nabla U\right) \tag{3.110}$$

It should be noted that, as well as the vector M defined by eqn (3.108), each other vector M_1,

$$M_1 = F(U, \Psi)\nabla U \times \nabla \Psi, \tag{3.111}$$

containing any arbitrary function $F(U, \Psi)$ also obeys eqn (3.108). The determination of vector M for the scattered field is appreciably easier than the traditional solution of the scattering problem. The components of this vector being known, the separation of their real and imaginary parts yields the fields H and E (eqn 3.107).

This notation proves to be important for the analysis of certain pulse scattered fields. Thus a linearly polarized plane field with the components E_y and H_x, travelling in the z-direction, may be described by means of scalar functions

$$U_1 = y + \mathrm{i}x, \quad \Psi_1 = z + ct. \tag{3.112}$$

Substitution of eqn (3.112) in eqn (3.109) yields $M_x = 1$, $M_y = -\mathrm{i}$; comparing this result with the definition (3.107), we obtain the dimensionless values of the components of the plane field: $E_y = -1$, $H_x = 1$. An arbitrary waveform of this field may be written, according to eqn (3.111), as

$$M_1 = F_1(\Psi_1)\nabla U_1 \times \nabla \Psi_1. \tag{3.113}$$

Proceeding in a similar fashion, one can find the representation of a diverging spherical wave, using the spherical coordinates ρ, θ, φ (Fig. 3.15). Thus, introducing the scalar functions

$$U_2 = \mathrm{e}^{-\mathrm{i}\varphi}\,\mathrm{tg}\left(\frac{\theta}{2}\right), \quad \Psi_2 = \rho - ct, \tag{3.114}$$

one can find the components of vector M for this field:

$$M_{2\theta} = -\frac{\mathrm{i}\mathrm{e}^{-\mathrm{i}\varphi}}{\rho(1+\cos\theta)}, \quad M_{2\varphi} = -\mathrm{i}M_{2\theta}, \quad M_{2\rho} = 0. \tag{3.115}$$

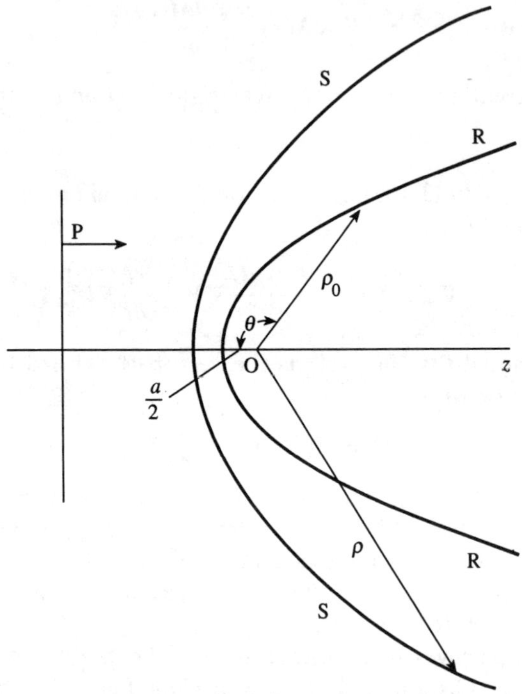

Fig. 3.15 Geometry of waveform-preserving reflection of a plane wave P to wave S on a parabolic reflector R; O, ρ and θ are the pole, radius vector, and polar angle of the parabolic surface R, respectively.

By analogy with eqn (3.113), vector M_2 can be presented in a form containing an arbitrary scalar function $F(\Psi_2)$:

$$M_2 = F_2(\Psi_2)\nabla U_2 \times \nabla \Psi_2. \tag{3.116}$$

We can now examine the scattering of an incident plane wave M_1 on some curvilinear surface R. Since eqn (3.108) governing the vector M_1, is linear, the linear combination

$$M = F_1(\Psi_1)\nabla U_1 \times \nabla \Psi_1 + CF_2(\Psi_2)\nabla U_2 \times \nabla \Psi_2 \tag{3.117}$$

representing the total field of the incident and scattered waves must also be a solution of this equation; the constant C in eqn (3.117) has to be defined from the boundary conditions at the scattering surface R.

Let us consider, using this approach, the scattering of an arbitrarily shaped plane wave P, incident on the convex side of a perfectly conducting axisymmetrical surface R along its axis of symmetry z (Fig. 3.15). Rewriting the scalar functions U_1 and Ψ_1 (eqn 3.112) in spherical coordinates,

$$U_1 = \rho \sin\theta \, e^{i\varphi}, \quad \Psi_1 = -\rho\cos\theta + ct, \tag{3.118}$$

and substituting eqn (3.118) in eqn (3.117), we can find the vector $\boldsymbol{M}$:

$$M_\rho = F_1 \sin\theta \, e^{i\varphi}, \tag{3.119}$$

$$M_\theta = F_1 \cos\theta \, e^{i\varphi} - \frac{iCF_2 e^{-i\varphi}}{\rho(1+\cos\theta)}, \tag{3.120}$$

$$M_\varphi = iF_1 e^{i\varphi} - \frac{CF_2 e^{-i\varphi}}{\rho(1+\cos\theta)}. \tag{3.121}$$

It is of interest to examine the case when an arbitrary waveform remains invariant in the course of scattering: $F_1 = F_2 = F$. Since the surface R is supposed to be perfectly conducting, the tangential component of the electric field, $E_\varphi = \operatorname{Im} M_\varphi$, vanishes on this surface:

$$\operatorname{Im} M_\varphi|_R = 0. \tag{3.122}$$

The boundary condition (3.122) can be fulfilled if the constant C is imaginary, $C = ia$, with

$$a = \rho_0(1+\cos\theta). \tag{3.123}$$

Here ρ_0 and θ are the polar coordinates of the curve forming the axisymmetric surface by rotation around the axis of symmetry z. The curve (3.123) is a parabola; its pole is located at the centre of the coordinate system (Fig. 3.15). Hence the surface R is a paraboloid.

The components of the EM field scattered by this paraboloid are determined directly from eqns (3.120)–(3.121):

$$\left.\begin{aligned} H_\theta = E_\varphi &= \frac{\rho_0}{\rho}\cos\varphi \, F(\rho - ct), \\ H_\varphi = E_\theta &= \frac{\rho_0}{\rho}\sin\varphi \, F(\rho - ct). \end{aligned}\right\} \tag{3.124}$$

Using these fields, one can write the radial component of the energy flow produced by scattered waveform as

$$P_\rho = \frac{c}{4\pi}\frac{a^2}{\rho^2(1+\cos\theta)^2}F^2(\rho - ct). \tag{3.125}$$

Thus we have obtained the time-domain description of the scattered wave. The main features of this unusual reflector are:

(a) the Pointing vector of the scattered field is directed radially: $P_\rho \neq 0$, $P_\theta = P_\varphi = 0$;

(b) this result is valid for an arbitrary waveform $F(\rho - ct)$;

(c) the preservation of the scattered waveform is probably a unique property of a paraboloid reflector.

This example shows the possibilities of the elegant time-domain method (Bateman 1955) in the problem of diffraction and scattering of transients, treated traditionally by computer simulations of inverse Fourier transforms.

3. In discussing the fundamentals of non-Fourier electrodynamics, one might ask whether it is a computational method only or whether it provides any new insight into EM pulse phenomena.

To clarify the characteristics of non-Fourier electrodynamics, let us make some important remarks:

a. The novelty of this subject is determined by the rapidly developing area of unusual problems associated with nonstationary fields excited by ultrashort and single-cycle waveforms in dispersive media. Some of these problems, e.g. reflection and refraction of such waveforms, are difficult to consider in the framework of the standard Fourier approach. Such waveform characteristics as shapes and time scales are considered only indirectly in the Fourier approach. In contrast, nonstationary electrodynamics, indicating the paramount role of these physical parameters, provides simple and explicit time-domain solutions of the said problems.

The special features of reflection–refraction processes for single-cycle waveforms at the interface of a dispersive medium are manifested especially clearly when the time scales of the waveform t_0 and medium T_0 are of the same order of magnitude. The cases $t_0 \ll T_0$ and $t_0 \gg T_0$ have been shown to relate to high-frequency and low-frequency asymptotes of reflectivity in the frequency domain (Fig. 3.3).

b. To find the link between the time-domain optics of single-cycle waveforms characterized by a small number of zero-crossing m ($m \approx 1$) and the frequency-domain optics of amplitude-modulated quasi-harmonic pulses with $m \gg 1$, e.g. eqn (1.15), one should use the model of Hermitian envelopes Φ_m. Formulae (1.33)–(1.34) for Φ_m show that the wings of high-order envelopes ($m \gg 1$) are similar to harmonic waves with carrier frequencies

$$\omega_0 = \begin{cases} \sqrt{2m}\, t_0^{-1}, & m = 2n, \\ \sqrt{m}\, t_0^{-1}, & m = 2n + 1. \end{cases} \tag{3.126}$$

Therefore Gaussian pulse envelopes containing many field oscillations ($m \gg 1$) bear a strong resemblance to the high-order Hermitian waveforms Φ_m. If parameter m is large enough, the difference between the time-domain and frequency-domain reflection coefficients of such pulses from the dispersive interface vanishes. Thus one can see the continuous transition between the Fourier and non-Fourier approaches.

Using this viewpoint, one can consider some well-known effects in traditional steady-state optics as limiting cases of relevant shape-dependent phenomena in nonstationary electrodynamics (see e.g. the Brewster effect for transients, Subsection 3.4.2).

c. The mathematical background elaborated for non-Fourier electrodynamics is based on new exact analytical nonseparable solutions of the Klein–Gordon and telegraph equations. This approach has been shown to simplify in principle the analysis of nonstationary fields by the reduction of differentiation and integration of these fields to subtraction and summation of finite quantities; moreover, the reflection–refraction problem becomes the solution of linear systems of algebraic equations. The development of this novel approach, reducing computer-memory requirements, represents a promising field in computational electromagnetics.

Thus nonstationary (non-Fourier) electrodynamics is characterized by its physical foundations and mathematical background. These propositions are distinguished from the relevant tenets of stationary (Fourier) electrodynamics. Nevertheless, both stationary and nonstationary electrodynamics, based on the same Maxwell equations, may be considered as two branches of one field of physics. Subject to the correlation of the time scales of waveform t_0 and dispersive medium T_0, their interaction may be attributed to either stationary ($t_0 \gg T_0$) or nonstationary ($t_0 \leqslant T_0$) electrodynamics. This duality promotes the use of both Fourier and non-Fourier approaches as complementary tools in the analysis of EM pulse processes. The choice of one or other method must be determined by the physics of the problem. Keeping in mind this obvious postulate, one can hope that the developing non-Fourier electrodynamics will not become a tempting dead end in the labyrinth of physics.

APPENDICES

A.1 The functions used in nonlinear geometrical optics

The solutions of the equations of nonlinear geometrical optics were shown in Subsection 1.3.1. to be expressed in terms the eigenfunctions of an axisymmetrical Laplace equation written in ellipsoidal coordinates (eqn 1.85). Some eigenfunctions of this equation used in expression (1.88) are given below:

$$\left.\begin{aligned} &P_0(\xi) = 1, \quad P_1(\xi) = \xi, \quad P_2(\xi) = \frac{1}{2}(3\xi^2 - 1),\\ &P_3(\xi) = \frac{1}{2}(5\xi^3 - 3\xi). \end{aligned}\right\} \tag{A.1.1}$$

$$\left.\begin{aligned} &Q_0(\xi) = \frac{1}{2}\ln\frac{1+\xi}{1-\xi}, \quad Q_1(\xi) = \frac{\xi}{2}\ln\frac{1+\xi}{1-\xi} - 1,\\ &Q_2(\xi) = \frac{1}{2}\left[\frac{3\xi^2 - 1}{2}\ln\frac{1+\xi}{1-\xi} - 3\xi\right],\\ &Q_3(\xi) = \frac{1}{2}\left[\frac{5\xi^3 - 3\xi}{2}\ln\frac{1+\xi}{1-\xi} - 5\xi^2 + \frac{4}{3}\right]. \end{aligned}\right\} \tag{A.1.2}$$

$$\left.\begin{aligned} &P_0(\mathrm{i}\varepsilon) = 1, \quad P_1(\mathrm{i}\varepsilon) = \varepsilon, \quad P_2(\mathrm{i}\varepsilon) = \frac{1}{2}(1 + 3\varepsilon^2),\\ &P_3(\mathrm{i}\varepsilon) = \frac{1}{2}(5\varepsilon^3 + 3\varepsilon). \end{aligned}\right\} \tag{A.1.3}$$

$$\left.\begin{aligned} &Q_0(\mathrm{i}\varepsilon) = \mathrm{arctg}\left(\frac{1}{\varepsilon}\right), \quad Q_1(\mathrm{i}\varepsilon) = \varepsilon\,\mathrm{arctg}\left(\frac{1}{\varepsilon}\right) - 1,\\ &Q_2(\mathrm{i}\varepsilon) = \frac{1}{2}\left[(1 + 3\varepsilon^2)\mathrm{arctg}\left(\frac{1}{\varepsilon}\right) - 3\varepsilon\right],\\ &Q_3(\mathrm{i}\varepsilon) = \frac{1}{2}\left[(5\varepsilon^3 + 3\varepsilon)\mathrm{arctg}\left(\frac{1}{\varepsilon}\right) - 5\varepsilon^2 - \frac{4}{3}\right]. \end{aligned}\right\} \tag{A.1.4}$$

The solutions of the modified axisymmetrical Laplace equation (1.85) contain the functions $P_n(\varepsilon)$, $Q_n(\varepsilon)$, obtained from $P_n(\xi)$ and $Q_n(\xi)$ by the replacement $\xi \to \varepsilon$.

A.2 Some properties of Bessel functions

$$J_{q-1}(x) - J_{q+1}(x) = 2\frac{\partial J_q(x)}{\partial x}. \tag{A.2.1}$$

$$J_{\pm q}(x)\big|_{|x|\gg 1} = \sqrt{\left(\frac{2}{\pi x}\right)}\left\{\cos\left(x \mp \frac{\pi}{2}q - \frac{\pi}{4}\right)\left[\sum_{k=0}^{q-1}\frac{(-1)^k}{(2x)^{2k}}\right.\right.$$

$$\left.\frac{\Gamma(q+2k+1/2)}{(2k)!\Gamma(q-2k+1/2)}\right] - \sin\left(x \mp \frac{\pi}{2}q - \frac{\pi}{4}\right)\left[\sum_{k=0}^{q-1}\frac{(-1)^k}{(2x)^{2k+1}}\right.$$

$$\left.\cdot\frac{\Gamma(q+2k+3/2)}{(2k+1)!\Gamma(q-2k-1/2)}\right]. \tag{A.2.2}$$

$J_q(x)$ is the Bessel function, Γ is the gamma function.

$$I_{p-1}(x) + I_{p+1}(x) = 2\frac{\partial I_p(x)}{\partial x}. \tag{A.2.3}$$

$$I_q(x)\big|_{x\gg 1} = \frac{e^x}{\sqrt{2\pi x}}\sum_{k=0}^{\infty}\frac{(-1)^k}{(2x)^k}\cdot\frac{\Gamma(q+k+1/2)}{k!\Gamma(q-k+1/2)}. \tag{A.2.4}$$

$$I_q(x)\big|_{x\ll 1} = \sum_{k=0}^{\infty}\frac{1}{k!\Gamma(q+k+1)}\cdot\left(\frac{x}{2}\right)^{q+2k}. \tag{A.2.5}$$

$I_q(x)$ is the modified Bessel function.

A.3 Table of matrix elements $P_{mq}(\alpha)$ and $Q_{mq}(\alpha)$ (eqn 3.21).

$$p_{mq} = P_{mq}\cdot 10^3, \quad q_{mq} = Q_{mq}\cdot 10^3$$

α	0.1	0.5	1	2	3	4	5	6	8	10
p_{01}	938	857	652	243	72	26	13	7	3	1
p_{03}	1	27	87	149	105	57	32	19	8	4
p_{05}	0	0	5	39	62	53	37	25	12	7
p_{07}	0	0	0	7	23	33	31	24	14	8
q_{01}	940	910	837	634	469	363	294	247	186	149
q_{03}	1	28	98	242	292	279	249	220	175	143
q_{05}	0	0	6	54	126	168	180	175	154	132
q_{07}	0	0	0	9	40	82	111	125	127	117

p_{12}	66	293	400	133	−70	−87	−68	−51	−31	−20
p_{14}	0	9	57	155	91	12	−18	−25	−22	−16
p_{16}	0	0	4	46	81	57	22	25	−10	−11
p_{18}	0	0	0	8	36	49	37	20	0	− 5
q_{12}	66	313	522	553	375	238	159	113	65	42
q_{14}	0	10	65	266	354	313	245	189	118	79
q_{16}	0	0	4	65	181	244	242	212	150	107
q_{18}	0	0	0	11	64	138	182	190	160	123

REFERENCES

Aaviksoo, J. and Kuhl, J. (1989). Transient reflection of light near excitonic resonances. *IEEE Journal of Quantum Electronics*, **25**, 2523–7.

Agrawal, G. P., Birman, J. L., Pattanayak, D. N., and Puri, A. (1982). Transient optical reflectivity from bounded non-local media. *Physical Review B*, **25**, 2715–29.

Akchmediev, N. N. and Korneev, V. I. (1987). Propagation of a periodic train of pulses in an optical fiber. *Radio Physics*, **30**, 1249–54 (in Russian).

Akchmediev, N. N., Korneev, V. I., and Mitskevich, N. V. (1988). N-modulational signals in a single-mode nonlinear optical fibre. *Zhurnal Eksperimental'noi i Teoreticheskoi Fiziki*, **94**, 159–70 (in Russian).

Bateman, H. (1955). *The mathematical analysis of electrical and optical wave motion on the basis of Maxwell equations*. Dover Publications, New York.

Belenov, E. M., Isakov, V. A., and Nazarkin, A. V. (1993). Non-resonant interaction of ultrashort electromagnetic pulses with multilevel quantum systems. *Kvantovaya Elektronika*, **20**, 1045–53 (in Russian).

Blachman, N. M. (1974). Sinusoids versus Walsh functions. *Proceedings of the IEEE*, **62**, 346–54.

Born, M. and Wolf, E. (1964). *Principles of optics*, Pergamon Press, Oxford.

Brillouin, L. (1960). *Wave propagation and group velocity*. Academic Press, New York.

Bui, M. D., Stuchly, S. S., and Gostache, G. I. (1991). Propagation of transients in dispersive dielectric media. *IEEE Transactions on Microwave Theory and Techiques*, **39**, 1165–71.

Campbell, I. H. and Fauchet, P. M. (1988). Temporal reshaping of ultra-short laser pulses after reflection from GaAs at Brewster's angle. *Optics Letters*, **13**, 634–6.

Chesnoy, J. and Mokhtari, A. (1987). Ultrafast oscillations of photoinduced absorption and birefringence observed by a technique of time-modulation. *Revue de Physique Appliqué*, **22**, 743–6.

Courant, R. and Hilbert, D. (1953). *Methods of mathematical physics*. Academic Press, New York.

Daubechies, I. (1988). Orthonormal bases of compactly supported wavelets. *Communications in Pure and Applied Mathematics*, **XLI**, 909–96.

Davies, R. W. (1954). The connection between the Smoluchowsky equation and the Kramers-Chandrasekhar equation. *Physical Review*, **93**, 1169–70.

Debye, P. (1929). *Polar molecules*. Dover Publications, New York.

Demtroder, W. (1982). *Laser spectroscopy*. Springer Verlag, Berlin.

Dodd, J. N. and Series, G. W. (1979). Time-resolved fluorescence spectroscopy. In *Progress in atomic spectroscopy*, (ed. W. Hanle and H. Kleinpopper), Plenum Press, New York.

Downey, P. M. and Tell, B. (1984). Picosecond photoconductivity studies of light-ion-bombarded InP. *Journal of Applied Physics*, **56**, 2672–4.

Dresselhaus, G., Kip, A.F., and Kittel, G. (1955). Plasma resonance in crystals: observations and theory. *Physical Review*, **100**, 618–25.

Durnin, J. (1987). Exact solutions for non-diffractive beams. *Journal of the Optical Society of America A*, **4**, 651–4.

D'yakov, Yu. E. (1993). Diffraction changes in the frequency spectrum of the propagating radiation (Wolf effect). *Kvantovaya Elektronika*, **20**, 1068–76 (in Russian).

Fattinger, Ch. and Grischkowsky, D. (1988). Point source terahertz optics. *Applied Physics Letters,* **53**, 1480–2.

Fattinger, Ch. and Grischkowsky, D. (1989). A Cherenkov source for freely-propagating terahertz beams. *IEEE Journal of Quantum Electronics*, **25**, 2608–10.

Fei, Z., Kivshar, Yu. S., and Vazquez, L. (1992). Resonant kink-impurity interactions in the Ψ^4 model. *Physical Review A*, **46**, 5214–20.

Feng, J. and Kneubuhl, F. (1993). Solitons in periodic structures with Kerr nonlinearity. *IEEE Journal of Quantum Electronics*, **29**, 590–8.

Feynman, R. P. (1972). *Statistical mechanics*. Benjamin, Massachusetts.

Frankel, M. Y., Gupta, S., Valdmanis, J. A., and Mourou, G. A. (1989). Picosecond pulse formation by transmission line discontinuities. *Electronics Letters*, **20**, 1363–5.

Frohlich, H. (1958) *Theory of dielectrics*, (2nd edn). Clarendon Press, Oxford.

Gabor, D. (1946). Theory of communications. *Journal of the Institution of Electrical Engineers*, **93**, 429–57.

Gibbs, H. M. (1985). *Optical bistability: controlling light by light*. Academic Press, New York.

Ginzburg, V. L. (1967). *Propagation of electromagnetic waves in plasma*. Pergamon Press, Oxford.

Grischkowsky, D. and Balant, A. C. (1982). Optical pulse compression based on enhanced frequency chirping. *Applied Physics Letters*, **41**, 1–3.

Guo, Y. and Uberall, H. (1994). The resonant effect of a wire in response to transient waves. *Journal of Electromagnetic Waves and Applications*, **8**, 355–66.

Harmuth, H. F. (1977). *Sequency theory*. Academic Press, New York.

Hasegawa, A. (1984). Generation of a train of soliton pulses by induced modulational instability in optical fibers. *Optics Letters*, **9**, 288.

Howell, W. E. and Uberall, H. (1990). Selective observation of resonances via their ringing in transient radar scattering, as illustrated for conducting and coated spheres. *IEEE Transactions on Antennas and Propagation*, **38**, 293–8.

Islam, M. N, Dijaili, S. P., and Gordon, J. P. (1988). Modulation instability-based laser interferometer switch near 1.5 μm. *Optics Letters*, **13**, 518–20.

Jackson, J. D. (1975). *Classical electrodynamics*, (2nd edn). Wiley, New York.

Janke, E., Emde, E., and Losch, F. (1960). *Tafeln höherer Funktionen*, (6th edn). Teubner, Stuttgart.

Jeunhomme, L. B. (1993). *Single-mode fiber optics*. Dekker, New York.

Joseph, R. M., Hagness., S. C., and Taflove, A. (1991). Direct time integration of Maxwell's equations in linear dispersive media with absorption for scattering and propagation of femtosecond electromagnetic pulses. *Optics Letters*, **16**, 1412–4.

Kabelka, V. and Mosolov, A. V. (1993). Determination of spectral width and chirp of ultrashort light pulses by analysis of second harmonic beam divergence. *Optics Communications*, **100**, 482–6.

Kaiser, G. (1994). *A friendly guide to wavelets*. Birkhauser, Boston.

Karlsson, M. and Hook, A. (1994). Fourth-order dispersion. *Optics Communications*, **104**, 303–7.

Kittel, C. (1948). The theory of ferromagnetic resonant absorption. *Physical Review*, **73**, 155–64.

Landau, L. D. and Lifshitz, E. M. (1984). *Electrodynamics of continuous media*. Pergamon Press, Oxford.

Lighthill, M. J. (1967). Non-linear evolution of solitary waves. *Proceedings of the Royal Society A*, **299**, 28–52.

Luebbers, R. J., Hunsberger, F., and Kunz, K. S. (1991). A frequency-dependent finite-difference time-domain foundation for transient propagation in plasma. *IEEE Transactions on Antennas and Propagation*, **39**, 29–34.

Ma, J. and Ciric, I. R. (1992). Transient response of a circular cylinder to an electromagnetic pulse. *Radio Science*, **27**, 561–7.

Martinet, R. K., Morlet, J., and Grossmann, A. (1987). Analysis of sound patterns through wavelet transforms. *International Journal of Pattern Recognition and Artificial Intelligence*, **1**, 273–302.

Mie, G. (1908). A contribution to the optics of turbid media, especially colloidal metallic suspensions. *Annals of Physics*, **25**, 377–445.

Mokhtari, A. A. and Chesnoy, J. (1989). Terahertz fluorescence quantum beams in a dye solution. *IEEE Journal of Quantum Electronics*, **25**, 2528–30.

Mollenauer, L. E., Stolen, R. and Islam, M. N. (1985). Experimental demonstration of soliton propagation in long fibers: loss compensated by Raman gain. *Optics Letters*, **10**, 229–31.

Morse, P. M. and Feshbach, H. (1953). *Methods of theoretical physics*. McGraw-Hill, New York.

Moseley, D. S. (1965). Non-separable solutions of the Helmholtz wave equation. *Quarterly of Applied Mathematics*, **22**, 354–7.

Newland, D. E. (1993). Harmonic wavelet analysis. *Proceedings of the Royal Society A*, **443**, 203–25.

Nuss, M. C. and Goossen, K. W. (1989). High-temperature superconductors with THz electrical pulses. *IEEE Journal of Quantum Electronics*, **25**, 2596–607.

Nussenzveig, H. M. (1993). *Diffraction effects in semiclassical scattering*. Academic Press, New York.

Onural, L. (1993). Diffraction from a wavelet point of view. *Optics Letters*, **18**, 846–8.

Oughstun, K. E. (1991). Pulse propagation in dispersive medium. *Proceedings of the IEEE*, **79**, 1376–90.

Paillette, M. (1987). Temperature-dependent behaviour of the Kerr constant in the vitreous state. *Journal of Non-Crystalline Solids*, **91**, 253–8.

Pang, L. Y., Kinttzer, E. S., and Fujimoto, J. G. (1991). Optical square wave generation in a diode-laser array *Optics Letters*, **16**, 1717–20.

Papazoglou, T. M. (1975). Transmission of a transient EM plane wave into a lossy half-space. *Journal of Applied Physics*, **46**, 3333–41.

Parrent, G. B., Jr and Skinner, T. J. (1961). Diffraction of a partially coherent light by a plane aperture *Optica Acta*, **8**, 93–100.

Pastol, Y., Arjavalingam, G., Halbout, J.-M., and Kopcsay, G.V. (1989). Coherent broadband microwave spectroscopy using picosecond optoelectronic antennas. *Applied Physics Letters*, **54**, 307–9.

Paulus, P., Brinker, W., and Jager, D. (1986). Generation of microwave pulses by optoelectronically switched resonators. *IEEE Journal of Quantum Electronics*, **22**, 108–11.

Putterman, S. I. (1974). *Superfluid hydrodynamics*. North Holland, Amsterdam.

Qian, Y. and Yamashita, E. (1991). Phase compensation and waveform reshaping of picosecond electrical pulses using dispersive microwave transmission lines. *IEEE Transactions on Microwave Theory and Techiques*, **39**, 924–9.

Rao, S. M. and Wilton, D. R. (1991). Transient scattering by conducting surfaces of arbitrary shape. *IEEE Transactions on Antennas and Propagation*, **39**, 56–61.

Ryzhik, I. M. and Gradstein, I. S. (1980). *Tables of integrals, series and products*. Academic Press, New York.

Sano, E., Nagatsuma, T., Shibata, T., and Iwata, A. (1989). Generation of picosecond electrical pulses by a pulse-forming optoelectronic device. *Applied Physics Letters*, **55**, 151–2.

Schoonaert, D. H. and Luypaert, P. J. (1973). Use of non-separable solutions of Helmholtz wave equation in waveguides and cavities. *Electronics Letters*, **9**, 617–18.

Schweber, S. S., Bethe, H. A., and de Hoffmann, F. (1955). *Mesons and fields*. Row, Peterson and Company, New York.

Scott, A. C. (1969). A non-linear Klein–Gordon equation. *American Journal of Physics*, **37**, 52–61.

Shen, H.-M., King, R. W. P., and Wu, T. T. (1990). New sensors for measuring very short electromagnetic pulses. *IEEE Transactions on Antennas and Propagation*, **38**, 838–45.

Sheng, Y., Roberge, D., and Szu, H. H. (1992). Optical wavelet transform, *Optical Engineering*, **31**, 1840–5.

Shvartsburg, A. B. (1974). Self-constriction of a wave packet in a non-linear medium. *Physics Letters A*, **48**, 257–9.

Shvartsburg, A. B. (1982). Non-linear dynamics and geometric properties of localized wave structures. *Physics Reports*, **83**, 109–49.

Shvartsburg, A. B. (1989). High speed resonant recording of frequency detuning in optical and submillimeter ranges using superfluidity phenomena. *Physics Letters*, **142**, 371–2.

Shvartsburg, A. B. (1993). *Non-linear pulses in integrated and waveguide optics*, Oxford University Press.

Shvartsburg, A. B. (1994). Non-Fourier plasma electrodynamics. *Physica Scripta*, **49**, 351–5.

Shvartsburg, A. B. (1995). Wavelet interactions with dispersive and lossy media. *SPIE Proceedings*, **2491**, 142–8.

Shvartsburg, A. B. and Stenflo, L. (1994). Reflection of wideband electromagnetic pulses at a plasma boundary. *Physica Scripta*, **49**, 712–4.

Spielmann, Ch., Kransz, F., Brabec, T., Wintner, E., and Schmidt, A. J. (1991). Femtosecond passive mode locking of a solid-state laser by a dispersively balanced non-linear interferometer. *Applied Physics Letters*, **58**, 2470–2.

Stegeman, G. I., Wright, E. M., Finlayson, N., Zanoni, R., and Seaton, C. T. (1988). Third-order non-linear integrated optics. *Journal of Lightwave Technology*, **6**, 953–70.

Stoyanov, D. V. and Dreischuh, T. N. (1992). On the use of non-Fourier eigensignals for measuring of short optical pulses by photomultiplier tubes. *Journal of Modern Optics*, **35**, 375–81.

Sun, E.-Y. and Rusch, W. V. T. (1994). Time-domain physical optics. *IEEE Transactions on Antennas and Propagation*, **42**, 9–15.

Szu, H., Sheng, Y. and Chen, J. (1992). The wavelet transform as a bank of the matched filters. *Applied Optics*, **31**, 3267–77.

Tai, K., Hasegawa, A., and Tomita, A. (1986a). Observation of modulational instability in optical fibers. *Physical Review Letters*, **56**, 135–8.

Tai, K., Jewell, J. L., and Tomita, A. (1986b). Generation of subpicosecond soliton-like optical pulses at 0.3 THz repetition rate by induced modulational instability. *Applied Physics Letters*, **49**, 236–8.

Tijhuis, A. G. (1984). Towards a stable marchihg-on-in-time method for two-dimensional transient electromagnetic scattering problems. *Radio Science*, **19**, 1311–17.

Tilton, H. B. (1986). *Waveforms. A modern guide to nonsinusoidal waves and nonlinear processes*. Prentice-Hall, Englewood Cliffs, New Jersey.

Tracy, E. R., Chen, H. H., and Lee, Y. C. (1984). Study of quasiperiodic solutions of the non-linear Schrodinger equation and modulational instability. *Physical Review Letters*, **53**, 218–21.

Trillo, S., Wabnitz, S., Wright, E. M., and Stegeman, G. I. (1989). *Optics Communications*, **70**, 166.

Usunov, I. M. (1990). Stationary periodical waves in optical fibers. *Optics Communications*, **79**, 23–5.

Wait, J. R. (1985). *Electromagnetic wave theory*. Harper and Row, New York.

Walsh, J. L. (1923). A closed set of normal orthogonal functions, *American Journal of Mathematics*, **55**, 5–24.

Weiner, A. M., Leaird, Reitze, D. H. and Paek, E. G. (1992). Spectral holography of shaped femtosecond pulses. *Optics Letters*, **17**, 224–6.

Wen, S. and Chi, S. (1989). Approximate solution of optical soliton with third-order dispersion, *Optical and Quantum Electronics*, **21**, 335–9.

Whitaker, J. F., Norris, T. B., Mourou, G. and Hsing, T. Y. (1987). Pulse dispersion and shaping in microstrip lines. *IEEE Transactions on Microwave Theory and Techniques*, **35**, 41–7.

Whitham, G. B. (1967). *Linear and non-linear waves*. Academic Press, New York.

Winful, H. G. (1985). Self-induced polarization changes in birefringent optical fibers. *Applied Physics Letters*, **47**, 213–15.

Wolf, E. (1986). Invariance of the spectrum of light on propagation. *Physical Review Letters*, **56**, 1370–4.

Zakharov, V. E., Manakov, S. V., Novikov, S. P., and Pitaevsky, L. P. (1980). *The theory of solitons* (in Russian). Nauka, Moscow.

Ziolkowski, R. W., Besieris, I. M. and Shaarawi, A. M. (1991). Localized wave representations of acoustic and electromagnetic radiation. *Proceedings of the IEEE*, **79**, 1371–8.

INDEX